Proceedings of the

Gökova Geometry-Topology Conference 2012

Editors

Selman Akbulut
Department of Mathematics, Michigan State University
East Lansing, MI 48824, USA

Denis Auroux
Department of Mathematics, University of California
Berkeley, CA 94720, USA

Turgut Önder
Department of Mathematics, Middle East Technical University
06531 Ankara, Turkey

GGT Gökova Geometry-Topology Conferences

International Press
www.intlpress.com

Published in 2013 by International Press, Somerville, Massachusetts, U.S.A. under license from Gökova Geometry-Topology Conferences.

ISBN 978-1-57146-270-1

Printed in the United States of America.

19th GÖKOVA GEOMETRY-TOPOLOGY CONFERENCE

May 28 – June 2, 2012

Participants of the 19th Gökova Geometry-Topology Conference

PREFACE

This year, in Gökova, we had many talks on topics of topology and geometry from the experts. Mirror symmetry, Floer homologies, exotic smooth structures on 4-manifolds, and fractals were some of the topics covered in this meeting. This volume contains interesting articles from this exciting nineteenth Gökova Conference. Special thanks goes to Cagatay Kutluhan, Jonathan Bloom, and Conan Leung for delivering mini-courses on Floer homology, Morse homology, and G_2 geometry, respectively. We thank all the participants for making this conference an exciting and stimulating event. We thank TÜBITAK and NSF for funding this conference, and thank IP for printing and distributing these proceedings. We also thank Selahi Durusoy for putting these proceedings together, and we finally thank Hotel Yücelen (which is located on the scenic shores of the Gökova Bay) for supporting and hosting this conference.

March 2013 — The Editors

CONTENTS

Proceedings of 19th Gökova
Geometry-Topology Conference
pp. 1 – 42

Lectures on the equivalence of Heegaard Floer and Seiberg–Witten Floer homologies

Çağatay Kutluhan

Abstract. This article gives a detailed account of the lectures delivered by the author on the construction, in joint work with Yi-Jen Lee and Clifford H. Taubes, of an isomorphism between Heegaard Floer and Seiberg–Witten Floer homologies of closed, connected, and oriented 3-manifolds.

1. Introduction

Over the last three decades, many breakthrough ideas have emerged from gauge theory and symplectic geometry with applications to topology. In the early 1980s, Donaldson revolutionized our understanding of the smooth topology of 4-manifolds using solutions of the anti-self-dual Yang–Mills equations [6]. In the late 1980s, Floer introduced the idea of generalizing Morse homology to certain infinite dimensional spaces. In this vein, Floer developed what came to be known as *Lagrangian Floer homology* [9] which associates to a pair of transversally intersecting Lagrangian submanifolds inside a symplectic manifold a relatively graded module. He also introduced an invariant of integer homology 3-spheres from Yang–Mills gauge theory, called *instanton Floer homology* [8], which he conjectured to be isomorphic to the Lagrangian Floer homology of the character varieties arising from a Heegaard decomposition. A more general version of this conjecture is known as the *Atiyah–Floer Conjecture*. Meanwhile, Donaldson introduced polynomial invariants of smooth 4-manifolds [7]. Instanton Floer homology carries Donaldson's invariants into a topological quantum field theory framework. In general, a topological quantum field theory (TQFT) in dimension n associates a module to every closed, oriented, and smooth n-manifold, and to every compact, oriented, and smooth $(n+1)$-manifold with boundary it assigns an element in the module associated to its boundary. These assignments are subject to certain axioms laid out by Atiyah [1]. As a consequence of these axioms, a TQFT associates to every closed, oriented, and smooth $(n+1)$-manifold an element of the ground ring. Even though the latter perspective has become useful in understanding the structure of Donaldson's invariants, these are still fairly hard to compute. This is why Donaldson's invariants were largely abandoned after the introduction of Seiberg–Witten equations in the mid 1990s [29].

The author is supported by a National Science Foundation Postdoctoral Research Fellowship under Award No. DMS-1103795.

Much like in Donaldson's theory, solutions of the Seiberg–Witten equations are used to define invariants of closed, oriented, and smooth 4-manifolds, called the *Seiberg–Witten invariants*, and a TQFT-like structure exists within Seiberg–Witten gauge theory as well. Namely, Seiberg–Witten Floer homology, developed by Kronheimer and Mrowka [17], associates to a given closed, connected, oriented, and smooth 3-manifold M three "graded" Abelian groups denoted by $\overline{HM}_*(M)$, $\widehat{HM}_*(M)$, and $\widetilde{HM}_*(M)$. These are invariants of smooth 3-manifolds, and they can be used to recapture, through homomorphisms induced by cobordisms, the Seiberg–Witten invariants of closed, oriented, and smooth 4-manifolds.

Even though the Seiberg–Witten invariants are more user friendly than Donaldson's invariants, these are also defined using geometric PDEs. Therefore, finding efficient ways to compute these invariants has been a challenging task. Twelve years ago, in order to give a differential topological description of the Seiberg–Witten invariants, and motivated by the Atiyah–Floer Conjecture, Ozsváth and Szabó developed Heegaard Floer homology [26, 27]. Heegaard Floer homology is an invariant of closed, connected, oriented, and smooth 3-manifolds which appears as a version of Lagrangian Floer homology arising from a Heegaard decomposition of the 3-manifold. Similar to Seiberg–Witten Floer homology, Heegaard Floer homology of a closed, connected, oriented, and smooth 3-manifold M has three flavors, denoted by $HF^\infty(M)$, $HF^-(M)$, and $HF^+(M)$, and these are used to define invariants of closed, oriented, and smooth 4-manifolds.

Despite their different origins, Heegaard Floer homology and Seiberg–Witten Floer homology have the same formal properties—such as the existence of a module structure over the ring $\mathbb{Z}[U] \otimes \wedge^*(H_1(M;\mathbb{Z})/torsion)$ and a long-exact sequence relating the three flavors—and they yield identical results where they can both be calculated; for example,

$$HF^\infty(S^3) \cong \mathbb{Z}[U,U^{-1}] \cong \overline{HM}_*(S^3), \quad HF^-(S^3) \cong \mathbb{Z}[U] \cong \widehat{HM}_*(S^3),$$

$$HF^+(S^3) \cong \mathbb{Z}[U,U^{-1}]/U\mathbb{Z}[U] \cong \widetilde{HM}_*(S^3).$$

Furthermore, the groups in each theory decompose into a direct sum over the set of isomorphism classes of Spin^c structures on M, which is a $H^2(M;\mathbb{Z})$-torsor. Ozsváth and Szabó conjectured that these two invariants are in fact the same. Our main result settles this conjecture:

Theorem 1.1 (Main Theorem in [18]). *Let M be a closed, connected, and oriented 3-manifold and $\mathfrak{s}$ be a Spin^c structure on M. Then, there exists a commutative diagram*

$$\begin{array}{ccccccccc}
\cdots \longrightarrow & HF^-(M,\mathfrak{s}) & \longrightarrow & HF^\infty(M,\mathfrak{s}) & \longrightarrow & HF^+(M,\mathfrak{s}) & \longrightarrow \cdots \\
 & \downarrow & & \downarrow & & \downarrow & \\
\cdots \longrightarrow & \widehat{HM}_*(M,\mathfrak{s},c_b) & \longrightarrow & \overline{HM}_*(M,\mathfrak{s},c_b) & \longrightarrow & \widetilde{HM}_*(M,\mathfrak{s},c_b) & \longrightarrow \cdots
\end{array}$$

where the top and the bottom rows are the standard long-exact sequences for Heegaard Floer homology and Seiberg–Witten Floer homology with balanced perturbations, while the vertical arrows are isomorphisms that preserve the relative gradings and intertwine the $\mathbb{Z}[U] \otimes \wedge^*(H_1(M;\mathbb{Z})/torsion)$-module structures.

The precise form of the conjectured relationship between Heegaard Floer and Seiberg–Witten Floer homologies, as stated in the above theorem, was first presented by Lee in [23]. Furthermore, she outlined a proof of the above theorem which essentially follows the same line of thought described in this article. The main difference, and the technical difficulty Lee ran into at the time, lies where one needs to handle the problems arising from the indefinite critical points of a self-indexing Morse function that is used to obtain a Heegaard decomposition of the 3-manifold.

The purpose of this expository article is to explain the content of the above theorem and to give an overview of its proof, which is joint work with Lee and Taubes, based on lectures delivered by the author at the 19th Gökova Geometry and Topology Conference. It is intended to serve as a guide for those who would like to read [19, 20, 21, 22] for more details. Our approach in the proof is to exploit the relationship between Seiberg–Witten theory and the theory of pseudo-holomorphic curves as established by Taubes. Roughly speaking, this relationship is based on the observation that, in a neighborhood of a pseudo-holomorphic curve, smooth sections of a particular line bundle belonging to solutions of the Seiberg–Witten equations are very close to being holomorphic and therefore their zero loci can be used to approximate the pseudo-holomorphic curve through some limit as the size of canonically chosen perturbations get large. That said, the construction of the isomorphisms in Theorem 1.1 requires the use of an auxiliary manifold Y and a particular type of geometry on it. The latter is used to define a twisted version of Hutchings's embedded contact homology (ECH), which we denote by *ech*. The proof of Theorem 1.1 is then broken into three parts. This is explained schematically as follows:

$$\begin{array}{ccc} \textit{ech} \text{ of } Y & \leftrightarrow & \text{Seiberg} - \text{Witten Floer cohomology of } Y \\ \updownarrow & & \updownarrow \\ \text{Heegaard Floer homology of } M & \cong & \text{Seiberg} - \text{Witten Floer homology of } M \end{array}$$

The organization of this paper is as follows: Section 2 describes the auxiliary manifold Y and its geometry. Section 3 defines *ech* and outlines the proof of its equivalence with Heegaard Floer homology. Section 4 describes a twisted version of Seiberg–Witten Floer cohomology and explains its equivalence with *ech*. Section 5 discuses a connected sum formula for Seiberg–Witten Floer homology and how it is used to relate the Seiberg–Witten Floer cohomology of Y with the Seiberg–Witten Floer homology of M. Finally, in Section 6, we conclude with some remarks.

Remark 1.2. Colin, Ghiggini, and Honda recently proved that ECH is isomorphic to $HF^+(-M)$ using open book decompositions [2, 3, 4, 5]. Together with Taubes's recent result on the equivalence of Seiberg–Witten Floer cohomology and ECH [34, 35, 36, 37, 38], their result provides an alternative way to see that $\widetilde{HM}_*(M)$ and $HF^+(M)$ are isomorphic. In fact, it might be possible to see, more directly, how the two approaches compare if one implements the construction described in the next section using a Heegaard diagram arising from an open book decomposition adapted to a contact structure. This way, one could try to relate the ECH and *ech* chain complexes in a natural way.

2. The auxiliary manifold and its geometry

Let M be a closed, connected, and oriented 3-manifold. We start by choosing a self-indexing Morse function $f : M \to \mathbb{R}$ with a single pair of definite critical points. In other words, f is a Morse function whose value at a critical point is equal to the index of that critical point. Such a function yields a Heegaard decomposition $M = U_0 \cup_\Sigma U_1$ where $U_0 = f^{-1}[0, 3/2]$ and $U_1 = f^{-1}[3/2, 3]$ are genus-$\textsc{g}$ handlebodies, and Σ is a closed, connected, and oriented surface of genus $\textsc{g}$ oriented as ∂U_0. Next, fix a gradient-like vector field for f, that is, a smooth vector field $\mathfrak{v}$ defined in the complement of the critical point set of f satisfying $df(\mathfrak{v}) > 0$. We rescale $\mathfrak{v}$ so as to satisfy $df(\mathfrak{v}) = 1$. The flow lines of $\mathfrak{v}$ that emanate from index-1 critical points of f intersect the surface Σ in $\textsc{g}$ disjoint simple closed curves $\mathbf{A} = \{A_1, \ldots, A_\textsc{g}\}$. Similarly, the flow lines of $-\mathfrak{v}$ that emanate from index-2 critical points of f intersect the surface Σ in $\textsc{g}$ disjoint simple closed curves $\mathbf{B} = \{B_1, \ldots, B_\textsc{g}\}$. We further choose $\mathfrak{v}$ so that $A_i \pitchfork B_j$ for each $i, j \in \{1, \ldots, \textsc{g}\}$. Then the ordered tuple $(\Sigma, \mathbf{A}, \mathbf{B})$, called a Heegaard diagram, determines the manifold M up to diffeomorphism.

A Heegaard diagram for M is part of the data that is required to define the Heegaard Floer homology groups. The remaining part of the data consists of a base point and a Spin^c structure. A base point is a fixed point $z \in \Sigma \setminus \bigcup(\mathbf{A} \cup \mathbf{B})$. A Heegaard diagram with a base point is called a *pointed Heegaard diagram.* The relevance of a pointed Heegaard diagram can be explained as follows: a Spin^c structure on M, in the sense of Turaev [40], is a nowhere vanishing vector field on M up to homotopy outside a 3-ball. Since a base point lies outside all the $\mathbf{A}$ and $\mathbf{B}$ curves, there exists a unique integral curve η_0 of $\mathfrak{v}$ that connects the definite critical points of f and intersects Σ transversally at this base point. On the other hand, any $\textsc{g}$-tuple of intersection points $\{x_1, \ldots, x_\textsc{g}\}$, where $x_i \in A_i \cap B_{\sigma(i)}$ for σ a permutation of $\{1, \ldots, \textsc{g}\}$, gives rise to a $\textsc{g}$-tuple of integral curves $\{\eta_1, \ldots, \eta_\textsc{g}\}$ of $\mathfrak{v}$ connecting pairs of indefinite critical points of f. One can change $\mathfrak{v}$ in a neighborhood of the curves $\eta_0, \eta_1 \ldots, \eta_\textsc{g}$ so that the resulting vector field is nowhere vanishing. In this way, one assigns a Spin^c structure to $\{x_1, \ldots, x_\textsc{g}\}$. As a result, the set of $\textsc{g}$-tuples of intersection points is partitioned into Spin^c equivalence classes.

Now, let $\mathfrak{s}$ be a Spin^c structure on M. The $\mathbf{A}$ and $\mathbf{B}$ curves divide the surface Σ into regions $\{\mathcal{D}_0, \mathcal{D}_1, \ldots, \mathcal{D}_N\}$ where $\mathcal{D}_0$ denotes the region containing the base point z. A 2-chain $\mathcal{P}$ that is an integer linear combination of $\mathcal{D}_1, \ldots, \mathcal{D}_N$ and whose boundary is an integer linear combination of $\mathbf{A}$ and $\mathbf{B}$ curves is called a *periodic domain.* Every periodic domain $\mathcal{P}$ determines an element $\dot{H}(\mathcal{P}) \in H_2(M; \mathbb{Z})$, and every element of $H_2(M; \mathbb{Z})$ can be represented by a periodic domain. A pointed Heegaard diagram $(\Sigma, \mathbf{A}, \mathbf{B}, z)$ is called *strongly $\mathfrak{s}$-admissible* if for every periodic domain $\mathcal{P}$ with $\langle c_1(\mathfrak{s}), H(\mathcal{P})\rangle = 2n \geq 0$, there exists a coefficient in $\mathcal{P}$ larger than n. This condition ensures that there are finitely many pseudo-holomorphic curves, up to translation, that enter into the definition of the Heegaard Floer differential. The following lemma claims that strong admissibility is equivalent to existence of a certain area form on Σ.

Lemma 2.1 (Lemma 1.1 in [19]). *A pointed Heegaard diagram* $(\Sigma, \mathbf{A}, \mathbf{B}, z)$ *is strongly* $\mathfrak{s}$*-admissible if and only if there exists an area form* w_Σ *on* Σ *such that*

- $\int_\Sigma w_\Sigma = 2$,
- $\int_{\mathcal{P}} w_\Sigma = \langle c_1(\mathfrak{s}), H(\mathcal{P})\rangle$ *for each periodic domain* $\mathcal{P}$.

Proof. The proof is similar to the proof of Lemma 4.12 in [26], except here we use $\mathfrak{s}$-renormalized periodic domains. See also the proof of the second bullet in Lemma 5.3 in [25]. □

Our aim in this section is to outline the construction, as in Sections 1 and 2 of [19], of an auxiliary manifold Y and a particular kind of geometry starting from the data provided by the strongly $\mathfrak{s}$-admissible Heegaard diagram $(\Sigma, \mathbf{A}, \mathbf{B}, z)$. We start by describing the manifold Y. First, fix a pairing Λ of index-1 and index-2 critical points of f, i.e., each $\mathfrak{p} \in \Lambda$ is a pair $\mathfrak{p} = (p_1, p_2)$ where p_i is an index-i critical point of f. Assume, without loss of generality, that the pairing respects the indexing of the $\mathbf{A}$ and $\mathbf{B}$ curves. Use p to denote the index-0 critical point of f, and use q to denote the index-3 critical point of f. The Morse Lemma tells us that, for some $\delta_* \in (0, c_0^{-1})$ where $c_0 >> 1$, there exist coordinate balls centered at index-1 and index-2 critical points of f on which f can be written respectively as

$$\begin{aligned} f &= 1 + x^2 + y^2 - 2z^2, \\ f &= 2 - x^2 - y^2 + 2z^2, \end{aligned}$$

where $(x^2 + y^2 + z^2)^{\frac{1}{2}} \leq 10\delta_*$. Denote by (r_+, θ_+, ϕ_+) and (r_-, θ_-, ϕ_-) the spherical coordinates in coordinate balls respectively centered at an index-1 and an index-2 critical point of f. Fix $R > -100 \ln \delta_*$, and for each $\mathfrak{p} = (p_1, p_2) \in \Lambda$, consider the spherical shell $\mathcal{H}_{\mathfrak{p}}$ centered at the origin with coordinates (r, θ, ϕ) defined by $e^{-R}(7\delta_*)^{-1} \leq r \leq e^R 7\delta_*$. After deleting the $r_+ < e^{-2R}(7\delta_*)^{-1}$ and $r_- < e^{-2R}(7\delta_*)^{-1}$ parts of the respective coordinate balls centered at p_1 and p_2, identify the respective $e^{-2R}(7\delta_*)^{-1} \leq r_+ \leq 7\delta_*$ and $e^{-2R}(7\delta_*)^{-1} \leq r_- \leq 7\delta_*$ parts with $\mathcal{H}_{\mathfrak{p}}$ via

$$\begin{aligned} (r_+, \theta_+, \phi_+) &= (e^{-R} r, \theta, \phi), \\ (r_-, \theta_-, \phi_-) &= (e^{-R} r^{-1}, \pi - \theta, \phi). \end{aligned} \tag{1}$$

Similarly, there exist coordinate balls centered at p and q on which f can be written respectively as

$$\begin{aligned} f &= x^2 + y^2 + z^2, \\ f &= 3 - x^2 - y^2 - z^2, \end{aligned}$$

where $(x^2 + y^2 + z^2)^{\frac{1}{2}} \leq 10\delta_*$. Denote by (r_+, θ_+, ϕ_+) and (r_-, θ_-, ϕ_-) the spherical coordinates in coordinate balls respectively centered at p and q. Consider the spherical shell $\mathcal{H}_0$ with coordinates (r, θ, ϕ) defined by $e^{-R}(7\delta_*)^{-1} \leq r \leq e^R 7\delta_*$ centered at the origin. After deleting the $r_+ < e^{-2R}(7\delta_*)^{-1}$ and $r_- < e^{-2R}(7\delta_*)^{-1}$ parts of the respective coordinate balls centered at p and q, identify the respective $e^{-2R}(7\delta_*)^{-1} \leq r_+ \leq 7\delta_*$ and

$e^{-2R}(7\delta_*)^{-1} \le r_- \le 7\delta_*$ parts with $\mathcal{H}_0$ via (1). The resulting manifold Y is diffeomorphic to $M\#_{\mathrm{G}+1}S^1 \times S^2$, the connected sum of M with $\mathrm{G}+1$ copies of $S^1 \times S^2$.

Next, we describe the particular type of geometry on Y that will enable us to establish the equivalence between Heegaard Floer and Seiberg–Witten Floer homologies.

2.1. A stable Hamiltonian structure on Y

Our aim is to construct a geometric structure on Y that will be consistent with the Heegaard Floer data on M. As Taubes has already established a relationship between Seiberg–Witten Floer cohomology and embedded contact homology, the appropriate sort of geometric structure here is a stable Hamiltonian structure. A stable Hamiltonian structure on Y is a pair (a, w) of a smooth 1-form a and a closed 2-form w on Y such that

- $da = \mathrm{H}w$ for a smooth function $\mathrm{H} : Y \to \mathbb{R}$,
- $a \wedge w$ is a volume form on Y.

On one end of a spectrum, contact structures are an example of stable Hamiltonian structures with a being a contact 1-form and $w = da$, while surface bundles over the circle are another example on the other end. The stable Hamiltonian structure we will describe on Y combines the two.

We start by defining the stable Hamiltonian structure on $\mathcal{H}_\mathfrak{p}$ for each $\mathfrak{p} \in \Lambda$. In order to do this, first introduce a new variable by $u = \ln r$ where $u \in [-2R+\ln 7\delta_*, 2R-\ln 7\delta_*]$. Then fix positive numbers $\delta < c_0^{-1}\delta_*$ and $\mathrm{x}_0 < \delta^3$. With δ and x_0 chosen, we may also need to increase the lower bound for the parameter R, namely, $R > -c_0 \ln \mathrm{x}_0$. Next, fix a smooth non-increasing function $\chi : \mathbb{R} \to [0,1]$ that equals 1 on $(-\infty, 0]$ and equals 0 on $[1, \infty)$. Also define the following three functions:

$$\mathrm{x}(u) = \mathrm{x}_0\chi(|u| - R - \ln\delta - 12),$$

$$\chi_+(u) = \chi(-u - \frac{R}{4}), \qquad \chi_-(u) = \chi(u - \frac{R}{4})$$

whose graphs are illustrated in Figure 1.

Now, consider the following contact 1-form on $\mathcal{H}_\mathfrak{p}$:

$$\mathrm{a} = (1 - 3\cos^2\theta)du - \sqrt{6}\cos\theta\sin^2\theta d\phi^1,$$

and define a 1-form on $\mathcal{H}_\mathfrak{p}$ by

$$a = \mathrm{x}(u)\mathrm{a} - 2\sqrt{6}(\chi_+ e^{2(u-R)} + \chi_- e^{-2(u+R)})\cos\theta\sin^2\theta d\phi + df_*, \tag{2}$$

where

$$f_* = (\chi_+ e^{2(u-R)} + \chi_- e^{-2(u+R)})(1 - 3\cos^2\theta).$$

With the choice of the parameters δ, x_0, and R as above, $a \wedge da$ is non-vanishing where $|u| \le R + \ln \mathrm{x}_0 - c_0$. Hence, we set $w = da$.

[1]This is the contact 1-form that appears on the boundary of a tubular neighborhood of a zero circle for a near-symplectic structure (see [32], also [12].)

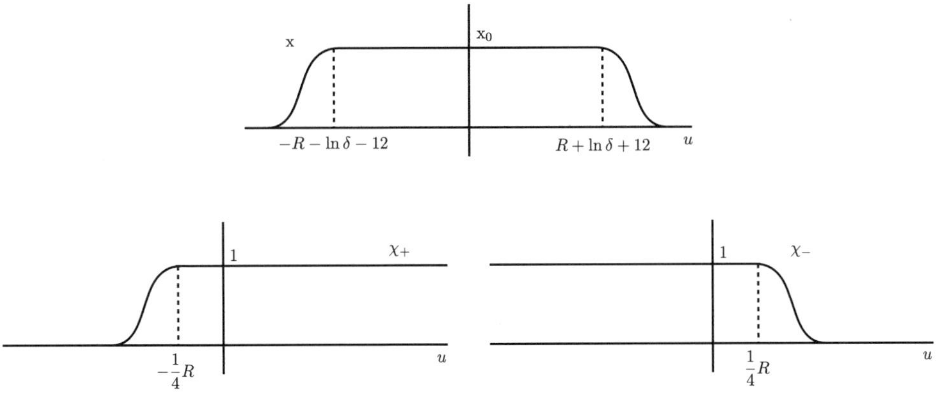

FIGURE 1. The functions x, χ_-, and χ_+.

In what follows, we describe the pair (a, w) on the complement in M of radius δ coordinate balls around all critical point of f. Denote the latter manifold by M_δ. For each index-1 or index-2 critical point of f, introduce new variables by $u_\pm = \ln r_\pm$. We will use an area form w_Σ on Σ, but with the opposite of the orientation of ∂U_0, supplied by Lemma 2.1 to construct the 2-form w on $f^{-1}([1,2]) \cap M_\delta$. To be more precise, use the integral curves of $\mathfrak{v}$ to identify $f^{-1}(1,2)$ with $(1,2) \times \Sigma$. Then pull back w_Σ via the projection from $[1,2] \times \Sigma$ onto Σ, and use the aforementioned identification to define a 2-form on $f^{-1}([1,2]) \cap M_\delta$. The latter agrees with w on the $\ln\delta < u_\pm < \ln 7\delta_*$ part of a coordinate ball around a given index-1 or index-2 critical point of f, therefore it extends w on $\bigcup_{\mathfrak{p}\in\Lambda} \mathcal{H}_\mathfrak{p}$ to the union of the latter with $f^{-1}([1,2]) \cap M_\delta$.

In order to extend the 1-form a to the union of $\bigcup_{\mathfrak{p}\in\Lambda} \mathcal{H}_\mathfrak{p}$ with $f^{-1}([1,2]) \cap M_\delta$, we need the following lemma.

Lemma 2.2 (Lemma 1.2 in [19]). *There exists a set of b_1, the first Betti number of M, points in $\Sigma \setminus \bigcup(\mathbf{A} \cup \mathbf{B})$ and a 1-form a_Σ defined on the complement of the base point z and these points such that*

- *$da_\Sigma = w_\Sigma$,*
- *The integral of a_Σ on each A_i or B_i is zero.*

Proof. We repeat the proof here for completeness. Consider the cohomology Mayer–Vietoris exact sequence for the Heegaard decomposition of M:

$$0 \to H^1(M;\mathbb{R}) \overset{\varphi}{\to} H^1(\Sigma;\mathbb{R}) \overset{\psi}{\to} Hom_\mathbb{Z}(F_{\mathbf{A}\cup\mathbf{B}}, \mathbb{R}) \overset{\delta}{\to} H^2(M;\mathbb{R}) \to 0,$$

where $F_{\mathbf{A}\cup\mathbf{B}}$ denotes the free $\mathbb{Z}$-module generated by $\mathbf{A}\cup\mathbf{B}$. The map φ is simply induced by the restriction of cochains, ψ is the map that sends the cohomology class of a closed 1-form on Σ to the hom-dual of $F_{\mathbf{A}\cup\mathbf{B}}$ defined by integrating this closed 1-form on curves in $\mathbf{A}\cup\mathbf{B}$, and $\delta(T)$ for $T \in Hom_\mathbb{Z}(F_{\mathbf{A}\cup\mathbf{B}}, \mathbb{R})$ is defined by $\delta(T)(H(\mathcal{P})) = T(\partial\mathcal{P})$. With the preceding understood, there exists a smooth 1-form a_z defined in the complement of the point z such that $w_\Sigma|_{\Sigma\setminus\{z\}} = da_z$, since $H^2(M \setminus \{z\}; \mathbb{R}) \cong 0$. Integrating a_z along curves in $\mathbf{A} \cup \mathbf{B}$ defines an element T_{a_z} of $Hom_{\mathbb{Z}}(F_{\mathbf{A}\cup\mathbf{B}}, \mathbb{R})$. If $\delta(T_{a_z}) = 0$, then by exactness of the above sequence, there exists a closed 1-form on Σ such that the integral of the sum of the latter and a_z vanishes on curves in $\mathbf{A} \cup \mathbf{B}$. Any given $(\mathrm{R}_1, \ldots, \mathrm{R}_N) \in \mathbb{R}^N$ gives rise to an element of $H^2(M; \mathbb{R})$ by sending a homology class represented by a periodic domain $\mathcal{P} = \sum_{i=1}^N c_i \mathcal{D}_i$ to $\sum_{i=1}^N c_i \mathrm{R}_i$. Therefore, we have a surjective homomorphism $\Phi : \mathbb{R}^N \to H^2(M; \mathbb{R})$. Now, choose a b_1-dimensional subspace of $\mathbb{R}^N$ whose restriction to Φ is an isomorphism. Fix points $\{z_i\}_{i=1,\ldots,b_1}$ in the interiors of the corresponding regions. Denote by R_i the integral of a_z on the boundary of the region $\mathcal{D}_i$. As a result, $\delta(T_{a_z}) = \Phi(\mathrm{R}_1, \ldots, \mathrm{R}_N)$, and we can modify a_z in $\Sigma \setminus \{z, z_1, \ldots, z_{b_1}\}$ so that the resulting 1-form a'_z has an associated $T_{a'_z} \in Hom_{\mathbb{Z}}(F_{\mathbf{A}\cup\mathbf{B}}, \mathbb{R})$ with $\delta(T_{a'_z}) = 0$. □

With the 1-form a_Σ in hand, we can find a 1-form $\hat{a}_\Sigma$ on Σ that is zero near the base point z and the set of b_1 points as in Lemma 2.2, that agrees with a_Σ outside a collection of disjoint disks with small area centered at the same points, and $d\hat{a}_\Sigma = \mathrm{H} w_\Sigma$ for some smooth function $\mathrm{H} : \Sigma \to \mathbb{R}$ that equals 1 on the complement of these disks. Then Part 7 in Section 1c of [19] explains how to extend the 1-form a to the union of $\bigcup_{\mathfrak{p}\in\Lambda} \mathcal{H}_{\mathfrak{p}}$ with $f^{-1}([1,2]) \cap M_\delta$ so that its restriction to $f^{-1}([1,2]) \cap M_\delta$ can be written as $\hat{a}_\Sigma + h(t)dt + d\varphi$ where t denotes the Euclidean coordinate on $[1,2]$, h is a function on $[1,2]$ that equals 1 near $t = 1$ and $t = 2$, and φ is a compactly supported t-dependent function on Σ. Furthermore, $da = \mathrm{H} w$ on $f^{-1}([1,2]) \cap M_\delta$.

In order to extend (a, w) to the remainder of Y, first extend (a, w) to the union of $\bigcup_{\mathfrak{p}\in\Lambda} \mathcal{H}_{\mathfrak{p}}$ with a neighborhood of $f^{-1}([1,2]) \cap M_\delta$. This can be done by identifying neighborhoods of $\Sigma_1 := f^{-1}(1) \cap M_\delta$ and $\Sigma_2 := f^{-1}(2) \cap M_\delta$ with $[1 - 2\delta^2, 1 + \delta^2] \times \Sigma_1$ and $[2 - \delta^2, 2 + 2\delta^2] \times \Sigma_2$ respectively via the integral curves of $\mathfrak{v}$, since the latter is tangent to the boundary of the radius δ coordinate balls centered at index-1 and index-2 critical points of f. With the preceding understood, w can be extended so as to be equal to w_Σ on these neighborhoods. On the other hand, a can be extended by Lie transport via the gradient-like vector field $\mathfrak{v}$. Then a construction similar to the extension of (a, w) to the union of $\bigcup_{\mathfrak{p}\in\Lambda} \mathcal{H}_{\mathfrak{p}}$ with $f^{-1}([1,2]) \cap M_\delta$ extends the pair to $(\bigcup_{\mathfrak{p}\in\Lambda} \mathcal{H}_{\mathfrak{p}}) \cup M_\delta$. Finally, (a, w) extends to the union of $(\bigcup_{\mathfrak{p}\in\Lambda} \mathcal{H}_{\mathfrak{p}}) \cup M_\delta$ with $\mathcal{H}_0$ so as to restrict on $\mathcal{H}_0$ as follows:

- $a = 2(\chi_+ e^{2(|u|-R)} + \chi_- e^{-2(|u|+R)})du + \hat{a}_0$,
- $w = \sin\theta d\theta \wedge d\phi$,

where $\hat{a}_0$ is a certain smooth 1-form on S^2. To be more explicit, the latter is equal to $\frac{1}{2\pi}\sin\theta d\theta \wedge d\phi$ outside disks of small radius centered at points that are identified with the $b_1 + 1$ points in Lemma 2.2 via the flow of $\mathfrak{v}$.

We orient the manifold Y so that $a \wedge w > 0$. Therefore, the inclusion of M_δ into Y is an orientation reversing embedding. By way of a summary, the stable Hamiltonian structure (a, w) and the associated Reeb vector field $\mathcal{R}$[2] have the following properties:

$$\left.\begin{array}{l}
\text{(a) On } \mathcal{H}_0,\ w = \sin\theta d\theta \wedge d\phi \text{ and } \mathcal{R} = \frac{1}{2(\chi_+ e^{2(u-R)} + \chi_- e^{-2(u+R)})}\frac{\partial}{\partial u}. \\
\text{(b) On } M_\delta,\ w \text{ restricts to each constant } f \text{ slice as an area form, and the vector} \\
\quad \text{field } \mathcal{R} \text{ on } M_\delta \text{ agrees with the gradient-like vector field } \mathfrak{v}. \\
\text{(c) Fix } \mathfrak{p} \in \Lambda. \text{ Then on } \mathcal{H}_\mathfrak{p}\text{:} \\
\quad - a = (\mathrm{x} + \mathfrak{g}')(1 - 3\cos^2\theta)du - \sqrt{6}\mathfrak{f}\cos\theta\sin^2\theta d\phi + 6\mathfrak{g}\cos\theta\sin\theta d\theta, \\
\quad - w = 6\mathrm{x}\cos\theta\sin\theta d\theta \wedge du - \sqrt{6}d(\mathfrak{f}\cos\theta\sin^2\theta d\phi), \\
\quad - \mathcal{R} = \frac{1}{(\mathfrak{f}(\mathrm{x}+\mathfrak{g})(1+9\cos^4\theta))^{\frac{1}{2}}}(\mathfrak{f}(1-3\cos^2\theta)\frac{\partial}{\partial u} - \sqrt{6}\mathrm{x}\cos\theta\frac{\partial}{\partial\phi} + \mathfrak{f}'\cos\theta\sin\theta\frac{\partial}{\partial\theta}), \\
\quad \text{where} \\
\qquad \mathfrak{f} = \mathrm{x} + 2(\chi_+ e^{2(u-R)} + \chi_- e^{-2(u+R)}), \qquad \mathfrak{g} = \chi_+ e^{2(u-R)} - \chi_- e^{-2(u+R)}. \\
\text{(d) Having written } H_2(Y;\mathbb{Z}) \cong H_2(M;\mathbb{Z}) \oplus H_2(\mathcal{H}_0;\mathbb{Z}) \oplus (\oplus_{\mathfrak{p}\in\Lambda} H_2(\mathcal{H}_\mathfrak{p};\mathbb{Z})), \text{ we} \\
\quad \text{have} \\
\quad - \langle [w], [S^2] \rangle = 2 \text{ for } [S^2] \text{ the positive generator of } H_2(\mathcal{H}_0;\mathbb{Z}), \\
\quad - \langle [w], [S^2] \rangle = 0 \text{ for } [S^2] \text{ the positive generator of } H_2(\mathcal{H}_\mathfrak{p};\mathbb{Z}) \text{ for some } \mathfrak{p} \in \Lambda, \\
\quad - \langle [w], F \rangle = \langle c_1(\mathfrak{s}), F \rangle \text{ for any } F \in H_2(M;\mathbb{Z}).
\end{array}\right\} \quad (3)$$

There is also a closed 1-form ν on Y which extends df on M_δ. To be more precise, the 1-form ν equals df on M_δ, it is given by $2(\chi_+ e^{2(|u|-R)} + \chi_- e^{-2(|u|+R)})$ on $\mathcal{H}_0$, and it is given by df_* on each $\mathcal{H}_\mathfrak{p}$. Furthermore, it satisfies $\nu \wedge w \geq 0$ with equality only where $u = 0$ and $1 - 3\cos^2\theta = 0$ on each $\mathcal{H}_\mathfrak{p}$. The 1-form ν is used to define an auxiliary 1-form $\hat{a}$ on Y as follows: let $\chi_\delta(u) = \chi(|u| - R - \ln\delta - 10)$. Then $\hat{a}$ is equal to ν on $M_\delta \cup \mathcal{H}_0$ and it is equal to $\chi_\delta a + (1 - \chi_\delta)\nu$ on each $\mathcal{H}_\mathfrak{p}$. The 1-form $\hat{a}$ satisfies $\hat{a}(\mathfrak{v}) = 1$ on M_δ and $\hat{a} \wedge w > 0$.

2.2. Closed integral curves for the stable Hamiltonian structure

Here, we characterize closed integral curves of $\mathcal{R}$ that are relevant to our story. We describe these curves via their intersections with various parts of the manifold Y. In this regard, let K^{-1} denote the oriented 2-plane field that is the kernel of a and is oriented by w. It follows from properties (a) and (d) in (3) that the Euler class $e_{K^{-1}}$ of K^{-1} satisfies:

- $\langle e_{K^{-1}}, [S^2] \rangle = 2$ where $[S^2]$ is the positive generator of $H_2(\mathcal{H}_0;\mathbb{Z})$,
- $\langle e_{K^{-1}}, [S^2] \rangle = -2$ where $[S^2]$ is the positive generator of $H_2(\mathcal{H}_\mathfrak{p};\mathbb{Z})$ for some $\mathfrak{p} \in \Lambda$.

There exists a unique $\Gamma \in H_1(Y;\mathbb{Z})$ and a Spin^c structure $\underline{\mathfrak{s}}$ on Y which satisfy $c_1(\underline{\mathfrak{s}}) = e_{K^{-1}} + 2PD(\Gamma)$, and

- $\langle PD(\Gamma), [S^2] \rangle = 0$ where $[S^2]$ is the positive generator of $H_2(\mathcal{H}_0;\mathbb{Z})$,
- $\langle PD(\Gamma), [S^2] \rangle = 1$ where $[S^2]$ is the positive generator of $H_2(\mathcal{H}_\mathfrak{p};\mathbb{Z})$ for some $\mathfrak{p} \in \Lambda$,
- The Spin^c structure $\underline{\mathfrak{s}}$ on Y restricts to the Spin^c structure $\mathfrak{s}$ on M.

[2]This is the vector field that spans the kernel of w and is normalized by a, i.e., $\iota_\mathcal{R} w = 0$ and $a(\mathcal{R}) = 1$.

Since $c_1(\underline{\mathfrak{s}}) = e_{K^{-1}} + 2PD(\Gamma)$, the second bullet implies that $\underline{\mathfrak{s}}$ restricts to the trivial Spin^c structure on each $\mathcal{H}_{\mathfrak{p}}$. This is an important fact that will come up naturally as we proceed.

With $\Gamma \in H_1(Y;\mathbb{Z})$ as above, let $\mathcal{Z}_{ech}$ denote the set of all finite collections Θ of pairs (γ_i, m_i) where γ_i is a closed integral curve of $\mathcal{R}$, and m_i is a positive integer such that $\Gamma = \sum_i m_i[\gamma_i]$. Additional constraints are required to determine the set of generators for the embedded contact homology chain complex. In order to explain these constraints, we consider the linearized return map of the Reeb flow around a closed integral curve of $\mathcal{R}$. Having fixed a closed integral curve γ of $\mathcal{R}$ and a point $x_0 \in \gamma$, the latter is a linear map $P_{\gamma,x_0} : K^{-1}|_{x_0} \to K^{-1}|_{x_0}$ which can be regarded as an element of $SL(2,\mathbb{R})$ when a trivialization of K^{-1} over γ is chosen. The curve γ is called *non-degenerate* if no power of P_{γ,x_0} has an eigenvalue equal to 1. For a non-degenerate closed integral curve γ, either both eigenvalues of P_{γ,x_0} are real, in which case γ is called *hyperbolic*, or both eigenvalues are on the unit circle, in which case γ is called *elliptic*. The constraint on a closed integral curve γ_i to be a part of a generator for the embedded contact homology chain complex is that it is non-degenerate, and the constraint on m_i is that $m_i = 1$ if γ_i is hyperbolic. We shall denote by $\mathcal{Z}_{ech,M}$ the subset of $\mathcal{Z}_{ech}$ with these additional constraints. An element of the set $\mathcal{Z}_{ech,M}$ is called an *admissible orbit set.*

We now characterize closed integral curves of $\mathcal{R}$ that belong to a collection from $\mathcal{Z}_{ech}$. First of all, such a closed integral curve γ of $\mathcal{R}$ would have empty intersection with $\mathcal{H}_0$. Here is why: if γ were to intersect $\mathcal{H}_0$, then it follows from property (a) in (3) that it would do so as $[-R - \ln\delta_*, R + \ln\delta_*] \times \{\text{point} \in S^2\}$, and therefore has positive intersection with every constant u cross-sectional 2-sphere of $\mathcal{H}_0$ oriented by w. As a result, $\langle PD(\Gamma), [S^2]\rangle > 0$ where $[S^2]$ is the positive generator of $H_2(\mathcal{H}_0;\mathbb{Z})$, contrary to what is said in the first bullet above. In fact, γ lies in the union of $f^{-1}(1,2)$ and $\bigcup_{\mathfrak{p}\in\Lambda}\mathcal{H}_{\mathfrak{p}}$ because otherwise γ would eventually have to intersect $\mathcal{H}_0$. Before we proceed, note that there exists a unique closed integral curve of $\mathcal{R}$ which intersects the Heegaard surface Σ at the base point z. We denote this closed integral curve by γ_z. It lies in $M_\delta \cup \mathcal{H}_0$ and intersects each level set of f as well as each constant u cross-sectional 2-sphere of $\mathcal{H}_0$ exactly once. The significance of the closed integral curve γ_z will be explained in the next section.

There are exactly two closed integral curves of $\mathcal{R}$ that sit entirely in $\mathcal{H}_{\mathfrak{p}}$ for each $\mathfrak{p} \in \Lambda$. These are denoted by $\gamma_{\mathfrak{p}}^+$ and $\gamma_{\mathfrak{p}}^-$, defined respectively by $u = 0$, $\cos\theta = \frac{1}{\sqrt{3}}$ and $u = 0$, $\cos\theta = -\frac{1}{\sqrt{3}}$ (see Figure 2). Both of these curves are hyperbolic and their associated linearized return maps have positive eigenvalues. On the other hand, if γ intersects $\mathcal{H}_{\mathfrak{p}}$ but does not lie entirely in it, then $\gamma \cap \mathcal{H}_{\mathfrak{p}}$ lies where $\cos^2\theta < \frac{1}{3}$, otherwise it would intersect $f^{-1}((0,1)\cup(2,3)) \cap M_\delta$ and eventually $\mathcal{H}_0$. Moreover, it follows from property (c) in (3) that γ intersects each constant u cross-sectional sphere of $\mathcal{H}_{\mathfrak{p}}$ exactly once positively, and either $\cos\theta = 0$ or $0 < \cos^2\theta < \frac{1}{3}$ along $\gamma \cap \mathcal{H}_{\mathfrak{p}}$. In fact, if θ_+ and θ_- denote the polar angle coordinates at which γ intersects respectively the boundaries of the radius δ_* coordinate balls centered at the index-1 and the index-2 critical points of f belonging to $\mathfrak{p}$,

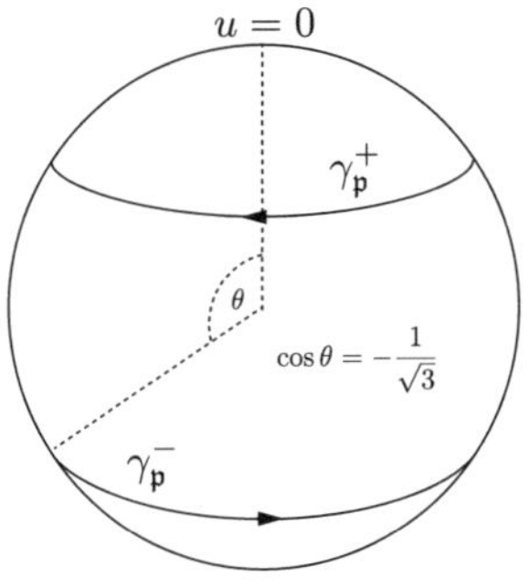

FIGURE 2. Integral curves that lie entirely in $\mathcal{H}_{\mathfrak{p}}$.

then $\theta_+ = \theta_-$; and the net change $\Delta\phi$ in value of the azimuthal angle coordinate as γ travels through the $|u| \leq R + \ln\delta_*$ part of $\mathcal{H}_{\mathfrak{p}}$ can be computed via the integral

$$\Delta\phi = -\sqrt{6}\int_{[-R-\ln\delta_*, R+\ln\delta_*]} \frac{\mathrm{x}(u)}{\mathfrak{f}(u)} \frac{\cos\theta(u)}{1 - 3\cos^2\theta(u)} du,$$

where $\theta(u)$ is the unique solution of the equation

$$\cos\theta \sin^2\theta = 2{\delta_*}^2 \cos\theta_\pm \sin^2\theta_\pm \frac{1}{\mathfrak{f}(u)}$$

with $1 - 3\cos^2\theta(u) > 0$.

Finally, we characterize the intersection of γ with M_δ. In this regard, let p_+ be an index-1 critical point and p_- be an index-2 critical point of f, and η denote a closed connected segment of γ that runs from the boundary of the radius δ coordinate ball centered at p_+ to the boundary of the radius δ coordinate ball centered at p_-. Suppose that $\eta \subset M_\delta$. Then η is in close proximity of an integral curve of $\mathfrak{v}$ that connects p_+ to p_-. The fundamental reason behind this is the fact that $\mathcal{R}$ agrees with $\mathfrak{v}$ in M_δ. With the preceding understood, a closed integral curve γ of $\mathcal{R}$ from a collection in $\mathcal{Z}_{ech}$ is broken into $2N$ closed connected segments $\{\eta^{\mathfrak{p}_1}, \eta_1, \ldots, \eta^{\mathfrak{p}_N}, \eta_N\}$ where $1 \leq N \leq \mathrm{G}$ and

- $\eta^{\mathfrak{p}_i} = \gamma \cap \mathcal{H}_{\mathfrak{p}_i}$ is a connected segment that starts from the index-2 critical point end of $\mathcal{H}_{\mathfrak{p}_i}$ and stops at the index-1 critical point end of $\mathcal{H}_{\mathfrak{p}_i}$,
- η_i is a connected segment that lies in M_δ. It starts from the index-1 critical point end of $\mathcal{H}_{\mathfrak{p}_i}$ and stops at the index-2 critical point end of $\mathcal{H}_{\mathfrak{p}_{i+1}}$ for $1 \leq i \leq N-1$, while η_N connects the index-1 critical point end of $\mathcal{H}_{\mathfrak{p}_N}$ to the index-2 critical point end of $\mathcal{H}_{\mathfrak{p}_1}$.

(see Figures 3 and 4.) The above observations bring up the following definition to be used in the upcoming structure theorem for elements of the set $\mathcal{Z}_{ech,M}$. An *index 1-2 cycle* is a non-empty, finite, cyclically order set $\{v_1, \ldots v_N\}$ consisting of closures of integral curves of $\mathfrak{v}$ that connect index-1 critical points of f to index-2 critical points of f in such a way that for $1 \leq i \leq N-1$ and the end point of v_i and the starting point of

υ_{i+1} as well as the end point of υ_N the starting point of υ_1 define distinct pairs form Λ. The above observations point out that any closed integral curve of $\mathcal{R}$ that belongs to a collection from the set $\mathcal{Z}_{ech,M}$ specifies an index 1-2 cycle. Conversely, we have the following statement:

Proposition 2.3 (Proposition 2.7 in [19]). *Given an index 1-2 cycle $\{\upsilon_1, \ldots \upsilon_N\}$, the set of closed integral curves of $\mathcal{R}$ that correspond to it enjoy a 1-1 correspondence with the set $\mathbb{Z}^N$. To be more precise, a closed integral curve of $\mathcal{R}$ corresponding to $\{\upsilon_1, \ldots \upsilon_N\}$ undergoes a change of $\Delta\phi_i + 2\pi m_i$, where $0 \leq \Delta\phi_i < 2\pi$ and $m_i \in \mathbb{Z}$, in its azimuthal angle coordinate as it travels through $\mathcal{H}_{\mathfrak{p}_i}$. Moreover, all of these closed integral curves are non-degenerate, hyperbolic, and the sign of the eigenvalues for the corresponding linearized return maps are $(-1)^N$ times the product of the orientation signs of the integral curves $\{\upsilon_1, \ldots \upsilon_N\}$. These signs are determined by comparing the orientation of Σ with the orientation determined by the* **A** *and* **B** *curves. Two pairs of an index 1-2 cycle and an N-tuple of integers specify the same closed integral curve if and only if they differ by an action of a cyclic permutation of $\{1, \ldots, N\}$ on the indices.*

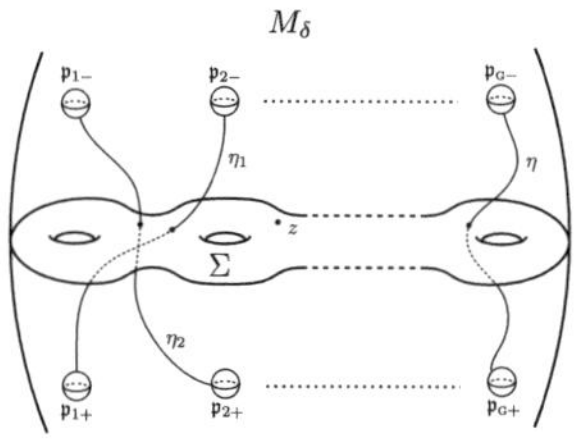

FIGURE 3. For each $i \in \{1, \ldots, \mathrm{G}\}$, the index-1 critical point $\mathfrak{p}_{i+}$ is paired with the index-2 critical point $\mathfrak{p}_{i-}$. Hence, the two segments on the left belong to a single closed integral curve, while the one on the right to another.

3. *ech* and its equivalence with Heegaard Floer homology

This section describes a twisted version of embedded contact homology for the stable Hamiltonian structure (a, w) on Y constructed from a strongly $\mathfrak{s}$-admissible Heegaard diagram $(\Sigma, \mathbf{A}, \mathbf{B}, z)$ arising from a self-indexing Morse function f on M as in the previous section, and justifies its equivalence with the Heegaard Floer homology of M.

3.1. The *ech* chain complex

The embedded contact homology chain complex is a free $\mathbb{Z}$-module generated by a set described as follows: Let Θ_0 be a 1-chain which represents Γ and does not intersect any closed integral curve from a collection in $\mathcal{Z}_{ech,M}$. Given $\Theta \in \mathcal{Z}_{ech,M}$, consider the relative homology classes of surfaces with boundary $\Theta - \Theta_0$. The latter is a

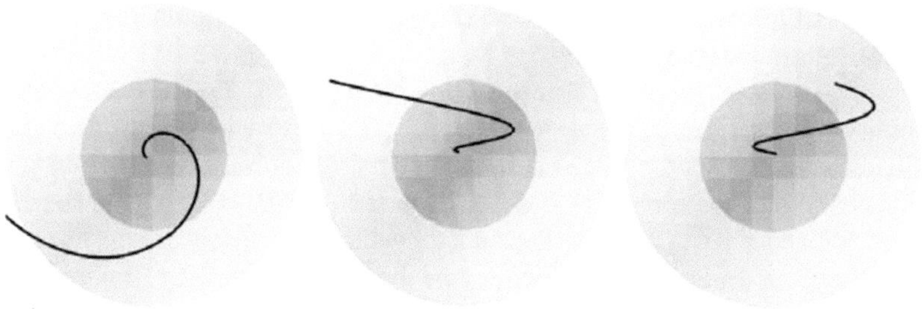

FIGURE 4. The intersection of a closed integral curve with a handle $\mathcal{H}_{\mathfrak{p}}$, as seen respectively (left to right) through the z-, y-, and x- axes.

$H_2(Y;\mathbb{Z})$-torsor, denoted by $H_2(Y,\Theta,\Theta_0)$. Next, consider pairs of the form (Θ, Z) consisting of $\Theta \in \mathcal{Z}_{ech,M}$ and $Z \in H_2(Y,\Theta,\Theta_0)$. Two such pairs (Θ, Z) and (Θ, Z') are deemed equivalent if $[Z - Z'] \cdot [\gamma_z] = 0$. We denote the set of equivalence classes under this relation by $\hat{\mathcal{Z}}_{ech,M}$. The latter admits an identification with $\mathcal{Z}_{ech,M} \times \mathbb{Z}$ via the surjection $H_2(Y,\Theta,\Theta_0) \to \mathbb{Z}$ sending Z to its algebraic intersection number with γ_z.

With the above understood, let ecc^{∞} denote the free $\mathbb{Z}$-module generated by $\hat{\mathcal{Z}}_{ech,M}$. We can define a submodule ecc^{-} as the one generated by the subset $\mathcal{Z}_{ech,M} \times \{\ldots, -2, -1\}$ and denote the quotient of ecc^{∞} by ecc^{-} by ecc^{+}. Then, there exists a natural short-exact sequence of free $\mathbb{Z}$-modules:

$$0 \longrightarrow ecc^{-} \xrightarrow{\ i\ } ecc^{\infty} \xrightarrow{\ \pi\ } ecc^{+} \longrightarrow 0. \tag{4}$$

The relative grading and the differential on these free $\mathbb{Z}$-modules are defined with the help of a suitable almost complex structure J on $\mathbb{R} \times Y$. The latter is an automorphism of $T(\mathbb{R} \times Y)$ whose square is the negative of the identity, and it is subject to additional constraints. In this regard, let s denote the Euclidean coordinate on $\mathbb{R}$. Then,

- $J\frac{\partial}{\partial s} = \mathcal{R}$.
- J is invariant under translations along the $\mathbb{R}$-factor.
- J preserves the kernel of the 1-form $\hat{a}$, and is compatible with the restriction of w on this plane field.
- J is invariant under translations of the azimuthal angle coordinate ϕ on each $\mathcal{H}_{\mathfrak{p}}$.
- Let

$$e_1 = -6\mathfrak{g}\cos\theta\sin\theta\frac{\partial}{\partial u} + (\mathrm{x}+\mathfrak{g}')(1-3\cos^2\theta)\frac{\partial}{\partial\theta},\quad e_2 = \frac{\partial}{\partial\phi} + \sqrt{6}\chi_\delta\mathfrak{f}\cos\theta\sin^2\theta\mathcal{R}.$$

Then $Je_1 = \sigma^{-1}e_2$ with σ a positive function of u and θ on each $\mathcal{H}_{\mathfrak{p}}$. (5)

A few remarks are in order. First of all, for any $v \in \mathrm{Ker}(a)$ we have $v - \hat{a}(v)\mathcal{R} \in \mathrm{Ker}(\hat{a})$. Hence, $w(\cdot, J\cdot)$ is positive definite on the kernel of a. Furthermore, it follows from the latter that there exists $r \geq 1$ such that the 2-form $\omega = ds \wedge a + rw$ tames J. On the other hand, J is compatible with the 2-form $\hat{\omega} = ds \wedge \hat{a} + w$. In particular, $\hat{\omega}(\cdot, J\cdot)$ defines an s-independent Riemannian metric on $\mathbb{R} \times Y$.

Having fixed an almost complex structure J on $\mathbb{R}\times Y$ satisfying the above constraints, a J-holomorphic curve is a smooth map $u : (S, j) \to (\mathbb{R}\times Y, J)$ where (S, j) is a (punctured) compact Riemann surface and $du \circ j = J \circ du$. A J-holomorphic curve is called *somewhere injective* if there exists a point $x \in S$ such that $u^{-1}(u(x)) = \{x\}$ and the linear map $du|_x : TS|_x \to T(\mathbb{R} \times Y)|_{u(x)}$ is injective. A somewhere injective curve is an embedding in the complement of finitely many points, and is determined by its image. On the other hand, a J-holomorphic curve is called *multiply covered* if it can be written as the composition of a somewhere injective curve $u : (S', j') \to (\mathbb{R} \times Y, J)$ and a holomorphic (branched) covering map $\varphi : (S, j) \to (S', j')$. In either case, the image C is a closed subset of $\mathbb{R} \times Y$ with finite 2-dimensional Hausdorff measure having no totally disconnected components, and the complement of finitely many points in C is a submanifold with J-invariant tangent space. We impose the following additional constraints on C:

- The integral of w over C is finite.
- There is an s-independent bound for the integral of $ds \wedge a$ over $C \cap [s, s+1] \times Y$ for any $s \in \mathbb{R}$.

The above two constraints together with the fact that ω tames J imply that constant s slices of C limit to a finite collection of closed integral curves of $\mathcal{R}$ as s tends to $\pm\infty$ (see [11] and [30]). To be more precise, there exists $s_0 \geq 1$ such that the $s > s_0$ part of C is a disjoint union of embedded cylinders on which the projection to the $\mathbb{R}$-factor has no critical points, similarly for the $s < -s_0$ part of C. Each such cylinder is called an *end* of C. It is called a *positive end* if it belongs to the $s > s_0$ part of C, and it is called a *negative end* if it belongs to the $s < -s_0$ part of C. Furthermore, constant s slices of each end of C converge pointwise to a closed integral curve of $\mathcal{R}$. With the preceding understood, we consider J-holomorphic curves whose images have ends at elements of $\mathcal{Z}_{ech,M}$. We will refer to such J-holomorphic curves as *admissible*. Hutchings defined the *ech* index for admissible J-holomorphic curves (see [13] and [14]). For a somewhere injective J-holomorphic curve, the *ech* index constitutes a lower bound for the Fredholm index of the associated deformation operator, and the two are equal if and only if the J-holomorphic curve is embedded. Given an admissible J-holomorphic curve whose image C has positive ends at $\Theta_+ \in \mathcal{Z}_{ech,M}$ and negative ends at $\Theta_- \in \mathcal{Z}_{ech,M}$, the *ech* index depends only on the relative homology class $[C] \in H_2(Y, \Theta_+, \Theta_-)$, and is denoted by $I([C])$. The *ech* index endows the *ech* chain complex with a relative $\mathbb{Z}$-grading that is well-defined modulo a non-negative integer p_M, the divisibility of $c_1(\mathfrak{s})$. For low *ech* index values Hutchings proved that admissible J-holomorphic curves consist of unions of embedded J-holomorphic curves together with possibly multiply covered $\mathbb{R}$-invariant cylinders, namely, $\mathbb{R} \times \Theta$ for $\Theta \in \mathcal{Z}_{ech,M}$. Therefore, from now on, we shall use the term J-holomorphic curve for finite collections of pairs $\{(C, n)\}$ where C is the image of a connected J-holomorphic curve in the above sense, and n is a positive integer indicating the covering multiplicity. The moduli space of J-holomorphic curves is endowed with the restriction of the weak*-topology on the space of currents. More precisely, given a J-holomorphic curve C, a neighborhood basis at C is indexed by pairs of the form (ϵ, ϖ) where ϵ is a positive real number and ϖ is a smooth 2-form on $\mathbb{R} \times Y$ such that a J-holomorphic curve C' is in one of these open neighborhoods if and only if

- $\sup_{x \in C} d(x, C') + \sup_{x \in C'} d(x, C) < \epsilon$,
- $| \int_C \varpi - \int_{C'} \varpi| < \epsilon$,

where $d(\cdot, \cdot)$ is the metric on $\mathbb{R} \times Y$ as defined by the Riemannian metric on $T(\mathbb{R} \times Y)$. There exists a version of Gromov Compactness for J-holomorphic curves in the current setting.

Theorem 3.1 (Proposition 5.5 in [19]). *Let $U \subset Y$ be an open set, $I \subset \mathbb{R}$ be a bounded, open interval, $\mathcal{L} \geq 1$, and $\{(C_k, n_k)\}$ be a sequence of J-holomorphic curves in $\mathbb{R} \times U$. Suppose that there exist $s \in \mathbb{R}$ and $\epsilon > 0$ such that*

$$\int_{C_k \cap [s-2, s+2] \times U_\epsilon} \hat{\omega} \leq \mathcal{L}$$

for each k, where U_ϵ denotes the set of points at a distance ϵ or less from the set U. Then, there exists a J-holomorphic curve $\{(C, n)\}$ and a subsequence of $\{(C_k, n_k)\}$ such that

- $\lim_{k \to \infty} (\sup_{x \in (\cup C) \cap I \times U} d(x, \cup C_k) + \sup_{x \in \cup C_k} d(x, (\cup C) \cap I \times U) = 0$,
- *Let ϖ be a smooth bounded 2-form defined in a neighborhood of $I \times U$. Then*

$$\lim_{k \to \infty} |\sum n \int_C \varpi - \sum n_k \int_{C_k} \varpi| = 0.$$

Note that translating a J-holomorphic curve along the $\mathbb{R}$-factor yields a J-holomorphic curve. This is because J is invariant under translations along the $\mathbb{R}$-factor. Therefore, the group $\mathbb{R}$ acts on the moduli space of J-holomorphic curves. This action is free away from the set of $\mathbb{R}$-invariant cylinders. Having said that, we now define the differential on the *ech* chain complex by describing the boundary of a generator $[\Theta_+, Z]$:

$$\partial^\infty_{ech}[\Theta_+, Z] = \sum_{\{\Theta_- \in \mathcal{Z}_{ech,M},\ W \in H_2(Y, \Theta_+, \Theta_-)\ |\ I(W) = 1\}} \#(\mathcal{M}(W)/\mathbb{R})[\Theta_-, Z - W]. \qquad (6)$$

The quantity $\#(\mathcal{M}(W)/\mathbb{R})$ refers to a signed count of admissible J-holomorphic curves (up to translation) representing the class W, which is finite thanks to the version of Gromov Compactness provided by Theorem 3.1. There is also a degree -2 chain map U on the *ech* chain complex that is defined by fixing a point $x \in \mathcal{H}_0$:

$$U[\Theta_+, Z] = \sum_{\{\Theta_- \in \mathcal{Z}_{ech,M},\ W \in H_2(Y, \Theta_+, \Theta_-)\ |\ I(W) = 2\}} \#\mathcal{M}_x(W)[\Theta_-, Z - W]. \qquad (7)$$

Here, $\mathcal{M}_x(W)$ denotes the moduli space of admissible J-holomorphic curves representing the class W that pass through the point $(0, x) \in \mathbb{R} \times Y$. Similarly, we can define the action of $\wedge^*(H_1(Y; \mathbb{Z})/torsion)$ by choosing a suitably generic basis of cycles that generate $H_1(Y; \mathbb{Z})/torsion$ and by a weighted sum of the algebraic intersection number of *ech* index-1 curves with these cycles in $\{0\} \times Y$.

Next we explain how admissible J-holomorphic curves foliate the complement in $\mathbb{R}\times Y$ of $\mathbb{R}\times\cup_{\mathfrak{p}\in\Lambda}(\gamma_{\mathfrak{p}}^{+}\cup\gamma_{\mathfrak{p}}^{-})$. This will be key to understanding the structure of the moduli space of J-holomorphic curves that enter, in particular, into the definition of the differential for the *ech* chain complex. We start by investigating admissible J-holomorphic curves that arise from level sets of the Morse function f and the handle $\mathcal{H}_0$. Since J preserves the kernel of $\hat{a}$ which restricts to df on M_δ, the intersection of level sets of f with M_δ are admissible J-holomorphic curves if $f\in(\delta^2,1-2\delta^2)\cup(1+\delta^2,2-\delta^2)\cup(2+2\delta^2,3-\delta^2)$, and are otherwise subsets of admissible J-holomorphic curves. These are described as follows (see Section 3(b) of [19]):

- A component $\mathcal{M}_0$ of the moduli space of admissible J-holomorphic curves consists of embedded 2-spheres and is $\mathbb{R}$-equivariantly diffeomorphic to $\mathbb{R}\times(-1,1)$. The part of $\mathcal{M}_0$ that is parametrized by $\mathbb{R}\times[-\frac{1}{2},\frac{1}{2}]$ consists of 2-spheres that arise from the constant u cross-sectional 2-spheres of $\mathcal{H}_0$ for $u\in[-R-\ln\delta,R+\ln\delta]$. The parts of $\mathcal{M}_0$ that are parametrized by $\mathbb{R}\times[\frac{1}{2},1)$ and $\mathbb{R}\times(-1,-\frac{1}{2}]$ respectively correspond to level sets of f for $f\in[\delta^2,1)$ and $f\in(2,3-\delta^2]$. Each admissible J-holomorphic curve from $\mathcal{M}_0$ has *ech* index 2.
- A component $\mathcal{M}_\Sigma$ of the moduli space of admissible J-holomorphic curves consists of embedded closed surfaces of genus G arising from the $f\in(1,2)$ level sets of f, and it is $\mathbb{R}$-equivariantly diffeomorphic to $\mathbb{R}\times(1,2)$. Each admissible J-holomorphic curve from $\mathcal{M}_\Sigma$ has *ech* index $2-2\mathrm{G}$.
- Two components, denoted by $\mathcal{M}_1$ and $\mathcal{M}_2$, of the moduli space of admissible J-holomorphic curves each consists of properly embedded 2-spheres with $2\mathrm{G}$ punctures, and each are $\mathbb{R}$-equivariantly diffeomorphic to $\mathbb{R}$. An admissible J-holomorphic curve from $\mathcal{M}_1$ projects onto M_δ as $f^{-1}(1)$, while an admissible J-holomorphic curve from $\mathcal{M}_2$ projects onto M_δ as $f^{-1}(2)$. These curves have $2\mathrm{G}$ negative ends each at a distinct $\gamma_{\mathfrak{p}}^{+}$ or $\gamma_{\mathfrak{p}}^{-}$, and no positive ends. Each admissible J-holomorphic curve from $\mathcal{M}_1$ or $\mathcal{M}_2$ has *ech* index $2-2\mathrm{G}$.

As for the admissible J-holomorphic curves contained in $\mathbb{R}\times\mathcal{H}_{\mathfrak{p}}$ for some $\mathfrak{p}\in\Lambda$, we have the following characterization:

- A component $\mathcal{M}_{\mathfrak{p}+}$ consists of embedded open disks that lie where $\cos\theta>\frac{1}{\sqrt{3}}$ and is $\mathbb{R}$-equivariantly diffeomorphic to $\mathbb{R}$. These curves have one positive end at $\gamma_{\mathfrak{p}}^{+}$, and no negative ends. They project onto $\mathcal{H}_{\mathfrak{p}}$ as the open disk defined by $u=0$ and $\cos\theta>\frac{1}{\sqrt{3}}$, and project onto $\mathbb{R}$ as the half infinite line $[s,\infty)$ for some $s\in\mathbb{R}$. Each admissible J-holomorphic curve from $\mathcal{M}_{\mathfrak{p}+}$ has *ech* index 1.
- A component $\mathcal{M}_{\mathfrak{p}-}$ consists of embedded open disks that lie where $\cos\theta<-\frac{1}{\sqrt{3}}$ and is $\mathbb{R}$-equivariantly diffeomorphic to $\mathbb{R}$. These curves have one positive end at $\gamma_{\mathfrak{p}}^{-}$, and no negative ends. They project onto $\mathcal{H}_{\mathfrak{p}}$ as the open disk defined by $u=0$ and $\cos\theta<-\frac{1}{\sqrt{3}}$, and project onto $\mathbb{R}$ as the half infinite line $[s,\infty)$ for some $s\in\mathbb{R}$. Each admissible J-holomorphic curve from $\mathcal{M}_{\mathfrak{p}+}$ has *ech* index 1.

- A component $\mathcal{M}_{\mathfrak{p}0}$ consists of embedded open annuli that lie where $\cos^2\theta < \frac{1}{3}$ and is $\mathbb{R}$-equivariantly diffeomorphic to $\mathbb{R}$. These curves have two positive ends at $\gamma_{\mathfrak{p}}^+$ and $\gamma_{\mathfrak{p}}^-$, and no negative ends. They project onto $\mathcal{H}_{\mathfrak{p}}$ as the open annulus defined by $u = 0$ and $\cos^2\theta < \frac{1}{3}$. Each admissible J-holomorphic curve from $\mathcal{M}_{\mathfrak{p}0}$ has *ech* index 0.

With the above understood, note that the only compact and connected admissible J-holomorphic curves are the ones contained in $\mathcal{M}_0$ and $\mathcal{M}_\Sigma$. This is because any such curve C has to intersect some curve from $\mathcal{M}_0$ or $\mathcal{M}_\Sigma$ nontrivially. Since J-holomorphic curves intersect with non-negative local intersection number and both C and the curve it intersects in $\mathcal{M}_0$ or $\mathcal{M}_\Sigma$ are compact, C has to intersect all curves from $\mathcal{M}_0$ or $\mathcal{M}_\Sigma$. But, this is possible only if the s-coordinate is unbounded on C, contradicting the assumption that C is compact.

Next, we make some preliminary observations about the admissible J-holomorphic curves that define the differential and various endomorphisms on the *ech* chain complex. First, we call an admissible J-holomorphic curve C an *ech-HF curve* if it does not contain any connected components from $\mathcal{M}_0$, $\mathcal{M}_\Sigma$, $\mathcal{M}_1$, $\mathcal{M}_2$, $\mathcal{M}_{\mathfrak{p}+}$, $\mathcal{M}_{\mathfrak{p}-}$, or $\mathcal{M}_{\mathfrak{p}0}$. An *ech-HF* curve C satisfies the following:

- C has empty intersection with $\mathbb{R} \times \mathcal{H}_0$ and with the parts of $\mathbb{R} \times M_\delta$ where $f \leq 1$ or $f \geq 2$.
- C has empty intersection with the part of $\mathbb{R} \times \mathcal{H}_{\mathfrak{p}}$ where $\cos^2\theta > \frac{1}{3}$ for each $\mathfrak{p} \in \Lambda$.
- A connected component of C intersecting $\mathbb{R} \times \mathcal{H}_{\mathfrak{p}}$ in the $\cos^2\theta = \frac{1}{3}$ locus is either the cylinder $\mathbb{R} \times \gamma_{\mathfrak{p}}^+$ or the cylinder $\mathbb{R} \times \gamma_{\mathfrak{p}}^-$.
- C has intersection number $\textsc{g}$ with each curve from $\mathcal{M}_\Sigma$.

By way of an explanation, if C were to intersect $\mathbb{R} \times \mathcal{H}_0$, the part of $\mathbb{R} \times M_\delta$ where $f \leq 1$ or $f \geq 2$, or the part of $\mathbb{R} \times \mathcal{H}_{\mathfrak{p}}$ where $\cos^2\theta > \frac{1}{3}$ for some $\mathfrak{p} \in \Lambda$, then it would have to intersect a curve from $\mathcal{M}_0$ since it does not contain any connected components from $\mathcal{M}_0$, $\mathcal{M}_1$, or $\mathcal{M}_2$. As a result, C must intersect each curve from $\mathcal{M}_0$, which is true even if C is non-compact due to the fact that the ends of C are disjoint from $\mathcal{H}_0$, the part of M_δ where $f \notin (1, 2)$, and the part of each $\mathcal{H}_{\mathfrak{p}}$ where $\cos^2\theta > \frac{1}{3}$. The latter reasoning also explains why C has empty intersection with the curves from $\mathcal{M}_0$ located at $\{s\} \times Y$ for sufficiently large values of $|s|$, contradicting the previous conclusion. On the other hand, suppose C has a connected component that intersects $\mathbb{R} \times \mathcal{H}_{\mathfrak{p}}$ where $\cos^2\theta = \frac{1}{3}$ and $u = 0$, and that it is neither $\mathbb{R} \times \gamma_{\mathfrak{p}}^+$ nor $\mathbb{R} \times \gamma_{\mathfrak{p}}^-$. Then it also intersects the part of $\mathbb{R} \times \mathcal{H}_{\mathfrak{p}}$ where $\cos^2\theta > \frac{1}{3}$, hence a curve from $\mathcal{M}_0$, $\mathcal{M}_{\mathfrak{p}+}$, or $\mathcal{M}_{\mathfrak{p}-}$. Since C does not contain any connected components from $\mathcal{M}_0$, $\mathcal{M}_{\mathfrak{p}+}$, or $\mathcal{M}_{\mathfrak{p}-}$, this possibility is ruled out by an argument similar to the one above. As for the claim in the last bullet, let C have positive ends at $\Theta_+ \in \mathcal{Z}_{ech,M}$ and negative ends at $\Theta_- \in \mathcal{Z}_{ech,M}$. Since the intersection of Θ_+ and Θ_- with M_δ each determines a $\textsc{g}$-tuple of integral curves of the gradient-like vector field $\mathfrak{v}$ which intersect Σ in $\textsc{g}$ distinct points, the part of C for sufficiently large values of $|s|$ intersects a unique curve from $\mathcal{M}_\Sigma$ in $\textsc{g}$ distinct points. Therefore, C has intersection number $\textsc{g}$ with every curve from $\mathcal{M}_\Sigma$ by the fact the intersection numbers are locally constant under small perturbations.

Finally, we describe the intersection of an *ech-HF* curve with $\mathbb{R} \times \mathcal{H}_{\mathfrak{p}}$ for any $\mathfrak{p} \in \Lambda$. Let C be an *ech-HF* curve and $\mathfrak{p} \in \Lambda$ be fixed.

$$\left.\begin{array}{l}\bullet \text{ If no end of } C \text{ is at } \gamma_{\mathfrak{p}}^{+} \text{ or } \gamma_{\mathfrak{p}}^{-}\text{, then } C \text{ has intersection number 1 with each curve from } \mathcal{M}_{\mathfrak{p}0}.\\ \bullet \text{ If one end of } C \text{ is at either } \gamma_{\mathfrak{p}}^{+} \text{ or } \gamma_{\mathfrak{p}}^{-}\text{, then } C \text{ has intersection number 1 with each curve from } \mathcal{M}_{\mathfrak{p}0} \text{ except for one, with which it has intersection number 0.}\\ \bullet \text{ If there are two ends of } C \text{ one of which is at } \gamma_{\mathfrak{p}}^{+} \text{ and the other is at } \gamma_{\mathfrak{p}}^{-}\text{, then } C \text{ has intersection number 1 with each curve from } \mathcal{M}_{\mathfrak{p}0} \text{ except for two, with which it has intersection number 0.}\end{array}\right\} \quad (8)$$

To sum up, the intersection of an *ech-HF* curve with a given $\mathbb{R} \times \mathcal{H}_{\mathfrak{p}}$ is a J-holomorphic strip with 0, 1, or 2 punctures.

Before we end this subsection, a few remarks are in order. First, since closed integral curves of $\mathcal{R}$ that comprise elements of $\mathcal{Z}_{ech,M}$ are all hyperbolic, there are no multiply covered admissible J-holomorphic curves in $\mathbb{R} \times Y$. Second, having fixed a point x in $\mathcal{H}_0$, Hutchings's characterization of the *ech* index-2 curves as unions of embedded J-holomorphic curves with possibly multiply covered $\mathbb{R}$-invariant cylinders lets us argue that the only admissible *ech* index-2 curve that passes through $(0, x) \in \mathbb{R} \times Y$ is the union of an embedded $\mathbb{R}$-invariant cylinder over an element of $\mathcal{Z}_{ech,M}$ and the unique embedded 2-sphere ${S^2}_x$ from $\mathcal{M}_0$ that passes through this point. As a result, the U-map in (7) can be written as

$$U[\Theta, Z] = [\Theta, Z - [{S^2}_x]].$$

3.2. The *ech* and Heegaard Floer homology equivalence

In order to establish the correspondence between Heegaard Floer homology and *ech*, we use Lipshitz's cylindrical reformulation of Heegaard Floer homology [25]. The latter exploits the same data provided by a strongly $\mathfrak{s}$-admissible Heegaard diagram $(\Sigma, \mathbf{A}, \mathbf{B}, z)$. The generators of the Heegaard Floer chain complex appear as pairs of the form (v, i) where $v = \{v_k = [1,2] \times \{x_k\}\}_{k=1}^{\mathrm{G}}$ with $x_k \in A_k \cap B_{\sigma(k)}$ for σ a permutation of $\{1, \ldots, \mathrm{G}\}$, and $i \in \mathbb{Z}$. Having identified $f^{-1}(1,2)$ with $(1,2) \times \Sigma$ via the gradient-like vector field $\mathfrak{v}$, each v_k corresponds to an integral curve of $\mathfrak{v}$ whose closure connects an index-1 critical point of f to an index-2 critical point of f. Therefore, if we decompose σ into cycles, each cycle corresponds to what we formerly called an index 1-2 cycle. Let us denote the set of generators for the Heegaard Floer chain complex by $\mathcal{Z}_{HF}$, and introduce a set with four elements $\mathrm{O} = \{0, 1, -1, \{1, -1\}\}$. Then, by Proposition 2.3, there exists a $1-1$ correspondence

$$\mathcal{Z}_{ech,M} \leftrightarrow \mathcal{Z}_{HF} \times \prod_{\mathfrak{p} \in \Lambda} (\mathbb{Z} \times \mathrm{O}). \quad (9)$$

The above correspondence is canonical when the $\mathbb{Z}$-factors on the right-hand side are considered as affine spaces over $\mathbb{Z}$. For each $\mathfrak{p} \in \Lambda$, a $\mathbb{Z}$-factor on the right-hand side is related to the net change in the azimuthal angle coordinate as a closed integral curve from a given element of $\mathcal{Z}_{ech,M}$ traverses $\mathcal{H}_{\mathfrak{p}}$, and an identification with $\mathbb{Z}$ is fixed once a lift of the azimuthal angle coordinate to $\mathbb{R}$ is chosen[3]; whereas the O-factor signifies whether a given element of $\mathcal{Z}_{ech,M}$ contains none of, exactly one of, or both of $\gamma_{\mathfrak{p}}^{+}$ and $\gamma_{\mathfrak{p}}^{-}$. The above correspondence lifts to a $1-1$ correspondence

$$\hat{\mathcal{Z}}_{ech,M} \leftrightarrow \mathcal{Z}_{HF} \times \mathbb{Z} \times \prod_{\mathfrak{p}\in\Lambda} (\mathbb{Z} \times \mathrm{O}), \tag{10}$$

where the identification of $\hat{\mathcal{Z}}_{ech,M}$ with $\mathcal{Z}_{ech,M} \times \mathbb{Z}$, as is explained in the previous subsection, is used.

The differential and various endomorphisms of the Heegaard Floer chain complex defined by Lipshitz require the choice of an almost complex structure on $\mathbb{R} \times [1,2] \times \Sigma$ subject to a number of constraints. Having identified $f^{-1}(1,2)$ with $(1,2) \times \Sigma$ via the gradient-like vector field $\mathfrak{v}$, these constraints translate as follows: J_{HF} be an almost complex structure on $\mathbb{R} \times f^{-1}(1,2)$, and s denote the Euclidean coordinate on the $\mathbb{R}$-factor. Then, we require

(11)

- $J_{HF}\frac{\partial}{\partial s} = \mathfrak{v}$.
- J_{HF} is invariant under translations along the $\mathbb{R}$-factor.
- J_{HF} preserves the kernel of the 1-form df, and it is compatible with the restriction of w on this plane field.
- J_{HF} is invariant under translations along integral curves of the gradient-like vector field $\mathfrak{v}$ near $\mathbb{R} \times (\mathfrak{A} \cup \mathfrak{D})$ where $\mathfrak{A}$ denotes the union of ascending manifolds of the index-1 critical points and $\mathfrak{D}$ denotes the union of descending manifolds of the index-2 critical points of f.

Note that the first three constraints in (11) agree on $\mathbb{R} \times (M_\delta \cap f^{-1}(1,2))$ with the first three constraints in (5). Moreover, these constraints indicate that J_{HF} is compatible with the symplectic form $ds \wedge df + w_\Sigma$, which restricts to $\hat{\omega}$ on $\mathbb{R} \times (M_\delta \cap f^{-1}(1,2))$. In addition to the above constraints, we require that

$$J_{HF}\frac{\partial}{\partial \phi_+} = \frac{\partial}{\partial \mathfrak{h}_+} \quad \text{and} \quad J_{HF}\frac{\partial}{\partial \phi_-} = \frac{\partial}{\partial \mathfrak{h}_-},$$

where

$$\mathfrak{h}_+ = 2e^{2u_+} \cos\theta_+ \sin^2\theta_+ \quad \text{and} \quad \mathfrak{h}_- = 2e^{2u_-} \cos\theta_- \sin^2\theta_-$$

on the intersection of $f^{-1}(1,2)$ and the radius $8\delta_*$ coordinate ball centered at respectively an index-1 and an index-2 critical point of f. Note also that this last constraint is consistent with the last two constraints listed above. With the preceding understood, Lipshitz considers J_{HF}-holomorphic curves S_0 in $\mathbb{R} \times (1,2) \times \Sigma$ satisfying the following properties:

- S_0 is the interior of a properly embedded surface S in $\mathbb{R} \times [1,2] \times \Sigma$ with 2G boundary components half of which lie on $\mathbb{R} \times \{1\} \times \mathbf{A}$, and the other half lie on $\mathbb{R} \times \{2\} \times \mathbf{B}$.

[3]However, there is no canonical lift. This is partly why the isomorphisms in Theorem 1.1 fail to be canonical.

- S is the complement of $2\textsc{g}$ distinct points in the boundary of a compact surface $\bar{S}$ with boundary. The s coordinate tends to $+\infty$ on sequences of points that limit to any one of $\textsc{g}$ of these points, while it tends to $-\infty$ on sequences of points that limit to any one of the remaining $\textsc{g}$ points.
- $\int_{S_0} w_\Sigma$ is finite, and there exists a constant $\kappa_S \geq 1$ such that for any $s \in \mathbb{R}$ we have $\int_{S_0 \cap([s,s+1]\times(1,2)\times\Sigma)} ds \wedge dt < \kappa_S$.
- The set of constant s slices of S converge pointwise to elements of $\mathcal{Z}_{HF}$ as s tends to $\pm\infty$.
- The associated deformation operator $D_S : L^2_1(N_S) \to L^2(N_S \otimes T^{0,1}S)$, where N_S denotes the normal bundle to S, has trivial cokernel.

The boundary of a given generator (v_+, i) of the Heegaard Floer chain complex is given by

$$\partial^\infty_{HF}(v_+, i) = \sum_{\{v_- \in \mathcal{Z}_{HF}\ [S]\in\pi_2(v_+,v_-)\ |\ ind(D_S)=1\}} \#\big(\mathcal{M}([S])/\mathbb{R}\big)(v_-, i - n_z([S])), \quad (12)$$

where $ind(D_S)$ denotes the Fredholm index of S, $\#\big(\mathcal{M}([S])/\mathbb{R}\big)$ is a signed count, modulo translations along the $\mathbb{R}$-factor, of J_{HF}-holomorphic curves in the relative homology class $[S]$, and $n_z([S])$ is the algebraic intersection number of S with $\mathbb{R} \times [1, 2] \times \{z\}$, which is non-negative. A degree -2 chain map U on the Heegaard Floer chain complex is defined by

$$U(v, i) = (v, i - 1). \quad (13)$$

Having established a correspondence between generators of the Heegaard Floer and *ech* chain complexes, we briefly explain how to relate the respective differentials. Given two generators $\Theta_+, \Theta_- \in \mathcal{Z}_{ech,M}$ which can be written respectively as $(v_+, (m_+, o_+))$ and $(v_-, (m_-, o_-))$ using the identification in (9), and a J_{HF}-holomorphic curve S of Fredholm index 1 that limits to v_+ at $+\infty$ and to v_- at $-\infty$, we first construct a canonical approximation $C_0 = \{C_{S0}, \{C_{\mathfrak{p}0}\}_{\mathfrak{p}\in\Lambda}\}$ to the desired admissible J-holomorphic curve C with positive ends at Θ_+ at $+\infty$ and negative ends at Θ_- at $-\infty$. Here, C_{S0} denotes a J-holomorphic curve with $2\textsc{g}$ boundary components in $\mathbb{R} \times f^{-1}(1, 2) \cap M_\delta$ and each $C_{\mathfrak{p}0}$ is a J-holomorphic curve with 2 boundary components in $\mathbb{R} \times \mathcal{H}_\mathfrak{p}$. The boundaries do not match, but there is a canonical $1-1$ correspondence between the boundary components of C_{S0} and the boundary components of $\bigcup_{\mathfrak{p}\in\Lambda} C_{\mathfrak{p}0}$. Moreover, the local boundary conditions on the deformation operators corresponding to C_{S0} and $C_{\mathfrak{p}0}s$ are coupled with respect to this correspondence. The curve C_{S0} looks very much like a J_{HF}-holomorphic curve of Fredholm index at most 1, while each $C_{\mathfrak{p}0}$ is one of the curves described in (8). As a matter of fact, C_{S0} arises as the image under the exponential map of a section of the bundle N_S. With the preceding understood, we construct a cobordism from the moduli space of such approximations to the component of the moduli space of admissible *ech* index 1 curves in $\mathbb{R} \times Y$. This cobordism maps to the interval $[0, 1]$ in a smooth and proper manner where the pre-image of 0 is the moduli space of approximations, while the pre-image of 1 is the component of the moduli space of admissible *ech* index 1 curves; and it induces an $\mathbb{R}$-equivariant diffeomorphism between the two spaces. To be more explicit, we have the following possibilities:

- C_{S0} has Fredholm index 1 and $(m_+, \mathrm{o}_+) = (m_-, \mathrm{o}_-)$,
- C_{S0} has Fredholm index 0, $(m_+, \mathrm{o}_+) \neq (m_-, \mathrm{o}_-)$, and either
  - $m_+ = m_-$ and $|\mathrm{o}_+| = |\mathrm{o}_-| + 1$, or
  - $m_+ = m_- \pm 1$ and $\mathrm{o}_+ = \mathrm{o}_-$.

Meanwhile, given a coherent system of orientations for the low index components of the moduli space of J_{HF}-holomorphic curves, there exists a canonical choice for a coherent system of orientations for the corresponding components of the moduli space admissible J-holomorphic curves. The latter is induced in part by canonical orientations on the moduli space of holomorphic curves in $\mathbb{R}\times\mathcal{H}_{\mathfrak{p}}$. Consequently, we get $\partial^{\infty}_{ech} = \partial^{\infty}_{HF} + \sum_{\mathfrak{p}\in\Lambda} \partial_*$ where ∂_* denotes the differential on the free $\mathbb{Z}$-module generated by $\mathbb{Z} \times \mathrm{O}$ defined as follows:

- $\partial_*(m, 0) = 0$ for each $m \in \mathbb{Z}$,
- $\partial_*(m, 1) = (m, 0) + (m + 1, 0)$ for each $m \in \mathbb{Z}$,
- $\partial_*(m, -1) = (m, 0) + (m - 1, 0)$ for each $m \in \mathbb{Z}$,
- $\partial_*(m, \{1, -1\}) = (m, -1) - (m, 1) + (m + 1, -1) - (m - 1, 1)$ for each $m \in \mathbb{Z}$.

As is easily checked, the homology of the chain complex $(\mathbb{Z}(\mathbb{Z} \times \mathrm{O}), \partial_*)$ is isomorphic to $\mathbb{Z} \oplus \mathbb{Z}$. Denote the latter graded group by $\hat{\mathrm{V}}$. The chains $(0, 0)$ and $(0, 1) - (1, -1)$ are closed and they represent two independent generators of the homology, the former in degree 0 and the latter in degree 1. Therefore,

Theorem 3.2 (Theorem 2.4 in [18]). *There exists a commutative diagram*

$$\begin{array}{ccccccccc}
\cdots \longrightarrow & ech^- & \longrightarrow & ech^{\infty} & \longrightarrow & ech^+ & \longrightarrow \cdots \\
& \downarrow & & \downarrow & & \downarrow & \\
\cdots \longrightarrow & HF^-(M, \mathfrak{s}) \otimes_{\mathbb{Z}} \hat{\mathrm{V}}^{\otimes \mathrm{G}} & \longrightarrow & HF^{\infty}(M, \mathfrak{s}) \otimes_{\mathbb{Z}} \hat{\mathrm{V}}^{\otimes \mathrm{G}} & \longrightarrow & HF^+(M, \mathfrak{s}) \otimes_{\mathbb{Z}} \hat{\mathrm{V}}^{\otimes \mathrm{G}} & \longrightarrow \cdots
\end{array}$$

where the top row is the long-exact sequence for ech associated to (4) and the bottom row is the standard long-exact sequence for the Heegaard Floer homology of M tensored with $\hat{\mathrm{V}}^{\otimes \mathrm{G}}$, while the vertical arrows are isomorphisms, induced by the correspondence (9), that preserve the relative gradings and intertwine the $\mathbb{Z}[U] \otimes \wedge^*(H_1(M;\mathbb{Z})/torsion)$-module structures.

For precise statements of the results summarized in this section that lead to a proof of the above theorem, see Theorem 1.1 in [20].

4. Seiberg–Witten Floer cohomology and its equivalence with *ech*

We start this section by reviewing some of the background in Seiberg–Witten Floer homology. We encourage the uninformed reader to refer to the book by Kronheimer and Mrowka [17] for an extensive treatment of the subject. After we cover some of the background material, we move on to describing a twisted version of Seiberg–Witten

Floer cohomology for certain Spin^c structures on Y as in [21]. Finally, we explain the proof of the equivalence of the latter with *ech*. This last part is along the same lines as Taubes's construction of a canonical isomorphism between ECH and Seiberg–Witten Floer cohomology [34, 35, 36, 37, 38], and the reader should refer to [21] for further details.

4.1. Preliminaries in Seiberg–Witten Floer homology

Here, we describe the Seiberg–Witten Floer homology groups of M. Start by fixing a Riemannian metric on M. Then a Spin^c structure $\mathfrak{s}$ on M consists of the following data:

- A rank-2 Hermitian bundle $\mathbb{S}$ over M, called *spinor bundle*,
- A Clifford multiplication, i.e., $\mathfrak{cl} : T^*M \to End_{\mathbb{C}}(\mathbb{S})$ an isometry onto the group of traceless, skew-Hermitian endomorphisms such that $\mathfrak{cl}(e^1)\mathfrak{cl}(e^2)\mathfrak{cl}(e^3) = 1_{\mathbb{S}}$ for any local oriented orthonormal basis $\{e^1, e^2, e^3\}$ for T^*M.

The first Chern class of the bundle $\mathbb{S}$ is denoted by $c_1(\mathfrak{s})$. With regard to the second part of the data, given $x \in M$ and any oriented orthonormal basis $\{e^1, e^2, e^3\}$ for T^*_xM, one can choose a frame for the fiber $\mathbb{S}_x$ so that

$$\mathfrak{cl}(e^1) = \begin{bmatrix} i & 0 \\ 0 & -i \end{bmatrix}, \quad \mathfrak{cl}(e^2) = \begin{bmatrix} 0 & -1 \\ 1 & 0 \end{bmatrix}, \quad \mathfrak{cl}(e^3) = \begin{bmatrix} 0 & i \\ i & 0 \end{bmatrix}.$$

Comparing the above definition to the definition of a Spin^c structure from Section 2, the orthogonal complement of a nowhere vanishing unit length vector field results in an oriented 2-plane field L on M which can be seen as the kernel of a unique nowhere vanishing smooth 1-form λ dual to that vector field. Then, the corresponding spinor bundle splits as $\mathbb{S}_L = \underline{\mathbb{C}} \oplus L$, and the Clifford multiplication is defined by

$$\mathfrak{cl}(\lambda) = \begin{bmatrix} i & 0 \\ 0 & -i \end{bmatrix},$$

while for any smooth unit length 1-form μ orthogonal to λ, $\mathfrak{cl}(\mu)(1,0) = (0, \mu^\dagger)$ where $\mu^\dagger$ denotes the vector field dual to μ. Conversely, the latter data defines a unique smooth unit length 1-form λ on M with $\mathfrak{cl}(\lambda) = \left[\begin{smallmatrix} i & 0 \\ 0 & -i \end{smallmatrix}\right]$, whose dual vector field is the nowhere vanishing vector field we started with. The set of isomorphism classes of Spin^c structures on M is a principle homogeneous space for $H^2(M;\mathbb{Z})$. The action of a class $e \in H^2(M;\mathbb{Z})$ on a Spin^c structure $\mathfrak{s}$ with spinor bundle $\mathbb{S}$ and Clifford multiplication $\mathfrak{cl}$ results in a Spin^c structure $\mathfrak{s} \otimes e$ with spinor bundle $\mathbb{S} \otimes E$ where E is a Hermitian line bundle over M with $c_1(E) = e$, and with Clifford multiplication $\mathfrak{cl} \otimes 1_E$.

Having fixed a Spin^c structure $\mathfrak{s}$ on M, consider the space $\mathcal{C}(M, \mathfrak{s})$ consisting of pairs of the form $(\mathbb{A}, \Psi)$ where $\mathbb{A}$ is a Spin^c connection, i.e., a Hermitian connection on $\mathbb{S}$ that satisfies the Leibniz rule

$$\nabla_{\mathbb{A}}(\mathfrak{cl}(\lambda)\Psi) = \mathfrak{cl}(\nabla\lambda)\Psi + \mathfrak{cl}(\lambda)(\nabla_{\mathbb{A}}\Psi),$$

with $\nabla\lambda$ denoting the covariant derivative with respect to the Levi-Civita connection, and Ψ is a smooth section of the spinor bundle $\mathbb{S}$. A Spin^c connection $\mathbb{A}$ on $\mathbb{S}$ induces a Hermitian connection $\mathbb{A}^t$ on the determinant bundle $det(\mathbb{S})$. Conversely, a Hermitian connection on $det(\mathbb{S})$ together with the Levi-Civita connection on TM induce a Spin^c connection on $\mathbb{S}$. The Lie group $\mathcal{G} = C^\infty(M, S^1)$ acts on $\mathcal{C}(M, \mathfrak{s})$ by the rule

$$\begin{aligned} \mathcal{G} \times \mathcal{C}(M, \mathfrak{s}) &\rightarrow \mathcal{C}(M, \mathfrak{s}) \\ (u, (\mathbb{A}, \Psi)) &\mapsto (\mathbb{A} - u^{-1}du \otimes 1_{\mathbb{S}}, u\Psi). \end{aligned}$$

A configuration $(\mathbb{A}, \Psi)$ is called *reducible* if $\Psi = 0$, otherwise it is called *irreducible.* Note that the action of $\mathcal{G}$ on $(\mathbb{A}, \Psi)$ has non-trivial stabilizer if and only if $(\mathbb{A}, \Psi)$ is reducible.

With the preceding understood, the *Seiberg–Witten equations* read:

$$\begin{aligned} *F_{\mathbb{A}^t} &= \Psi^\dagger \tau \Psi, \\ \mathcal{D}_{\mathbb{A}}\Psi &= 0. \end{aligned} \tag{14}$$

Here, $*F_{\mathbb{A}^t}$ is the Hodge dual of the curvature form of $\mathbb{A}^t$, $\Psi^\dagger\tau\Psi : T^*M \rightarrow i\mathbb{R}$ is defined by $\Psi^\dagger\tau\Psi(\lambda) = \Psi^\dagger \mathfrak{cl}(\lambda)\Psi$, and $\mathcal{D}_{\mathbb{A}}$ is the Dirac operator defined to be the composition

$$C^\infty(M, \mathbb{S}) \xrightarrow{\nabla_{\mathbb{A}}} C^\infty(M, T^*M \otimes \mathbb{S}) \xrightarrow{\mathfrak{cl}} C^\infty(M, \mathbb{S}).$$

The Seiberg–Witten equations arise as the variational equations of the *Chern–Simons–Dirac functional.* Having fixed a base Spin^c connection $\mathbb{A}_0$ on $\mathbb{S}$, the latter can be written as

$$\mathfrak{csd}(\mathbb{A}, \Psi) = -\frac{1}{8}\int_M (\mathbb{A}^t - \mathbb{A}_0{}^t) \wedge (F_{\mathbb{A}^t} + F_{\mathbb{A}_0{}^t}) + \frac{1}{2}\int_M \langle \mathcal{D}_{\mathbb{A}}\Psi, \Psi\rangle. \tag{15}$$

Note that the moduli space of solutions to the Seiberg–Witten equations is invariant under the action of $\mathcal{G}$, while the value of the Chern–Simons–Dirac functional changes as follows:

$$\mathfrak{csd}(u \cdot (\mathbb{A}, \Psi)) - \mathfrak{csd}(\mathbb{A}, \Psi) = 2\pi^2(c_1(\mathfrak{s}) \cup [u]) \cap [M],$$

for every $u \in \mathcal{G}$, where $[u] \in H^1(M; \mathbb{Z})$ is represented by $-\frac{i}{2\pi}u^{-1}du$. In particular, the Chern–Simons–Dirac functional is gauge invariant if and only if $c_1(\mathfrak{s})$ is torsion.

One can perturb the Seiberg–Witten equations using a closed 2-form ϖ:

$$\begin{aligned} *F_{\mathbb{A}^t} &= \Psi^\dagger \tau \Psi + i * \varpi, \\ \mathcal{D}_{\mathbb{A}}\Psi &= 0. \end{aligned} \tag{16}$$

The above perturbed Seiberg–Witten equations are obtained from a perturbed version of the Chern–Simons–Dirac functional defined by

$$\mathfrak{csd}_\varpi(\mathbb{A}, \Psi) = \mathfrak{csd}_\varpi(\mathbb{A}, \Psi) + \frac{1}{4}\int_M (\mathbb{A}^t - \mathbb{A}_0{}^t) \wedge i\varpi. \tag{17}$$

The perturbed Chern–Simons–Dirac functional changes under the action of $\mathcal{G}$ as follows:

$$\mathfrak{csd}_\varpi(u \cdot (\mathbb{A}, \Psi)) - \mathfrak{csd}_\varpi(\mathbb{A}, \Psi) = 2\pi^2((c_1(\mathfrak{s}) + \frac{1}{2\pi}[\varpi]) \cup [u]) \cap [M].$$

The cohomology class $c = 2\pi^2(c_1(\mathfrak{s}) + \frac{1}{2\pi}[\varpi]) \in H^2(M; \mathbb{R})$ is called the *period class.* Note that if the period class is non-zero, then the perturbed Seiberg–Witten equations in (16) have no reducible solutions. If ϖ is an exact 2-form, then the resulting perturbation is called an *exact* perturbation, otherwise it is called a *non-exact* perturbation. A non-exact perturbation is called

- *Balanced* if c is zero.
- *Positively/Negatively monotone* if $c = \kappa 2\pi^2 c_1(\mathfrak{s})$ for some $\kappa > 0$ or $\kappa < 0$ respectively.

In order to define the Seiberg–Witten Floer chain complexes, Kronheimer and Mrowka *blow up* (see Section of [17]) the quotient $\mathcal{B}(M,\mathfrak{s}) = \mathcal{C}(M,\mathfrak{s})/\mathcal{G}$ along the singular locus consisting of gauge equivalence classes of reducible solutions. Note that the latter set is non-empty if and only if the period class is zero. In this case, the resulting Banach manifold $\mathcal{B}^\sigma(M,\mathfrak{s})$ with boundary admits a vector field induced by the L^2-gradient of the perturbed Chern–Simons–Dirac functional in (17). This vector field is tangent to the boundary of $\mathcal{B}^\sigma(M,\mathfrak{s})$ and its critical points are pre-images of gauge equivalence classes of solutions to the equations in (16) under the blow-down map. The critical points that map to gauge equivalence classes of reducible solutions to the equations in (16) fall into two categories depending on whether or not the outward normal direction at a critical point on the boundary is contained in the corresponding stable manifold. These are called *boundary stable* and *boundary unstable*, respectively. The Seiberg–Witten Floer chain complexes $\overline{CM}_*(M,\mathfrak{s},c_{[\varpi]})$, $\widehat{CM}_*(M,\mathfrak{s},c_{[\varpi]})$, and $\widecheck{CM}_*(M,\mathfrak{s},c_{[\varpi]})$ are the free $\mathbb{Z}$-modules generated respectively by boundary stable and boundary unstable critical points, irreducible and boundary unstable critical points, and irreducible and boundary stable critical points; the respective differentials $\bar{\partial}$, $\hat{\partial}$, and $\check{\partial}$ are defined via a suitable count of trajectories of the aforementioned vector field. Here, $c_{[\varpi]}$ indicates the period class of the perturbation defined by the closed 2-form ϖ. In order to ensure transversality, one may need to use additional abstract perturbations. We will elaborate on this matter in the next subsection. The homologies of the chain complexes $(\overline{CM}_*(M,\mathfrak{s},c_{[\varpi]}),\bar{\partial})$, $(\widehat{CM}_*(M,\mathfrak{s},c_{[\varpi]}),\hat{\partial})$, and $(\widecheck{CM}_*(M,\mathfrak{s},c_{[\varpi]}),\check{\partial})$ are respectively denoted by $\overline{HM}_*(M,\mathfrak{s},c_{[\varpi]})$, $\widehat{HM}_*(M,\mathfrak{s},c_{[\varpi]})$, and $\widecheck{HM}_*(M,\mathfrak{s},c_{[\varpi]})$. Each of these groups is relatively graded by $\mathbb{Z}/p_M$ where p_M denotes the divisibility of $c_1(\mathfrak{s})$, each has the structure of a $\mathbb{Z}(U)\otimes\wedge^*(H_1(M;\mathbb{Z})/torsion)$-module, and they fit into a long exact sequence

$$\cdots \to \widehat{HM}_* \xrightarrow{p_*} \overline{HM}_* \xrightarrow{i_*} \widecheck{HM}_* \xrightarrow{j_*} \widehat{HM}_* \to \cdots$$

for each $(M,\mathfrak{s},c_{[\varpi]})$, analogous to the homology long exact sequence for a pair of a manifold and its boundary.

4.2. A twisted Seiberg–Witten Floer cohomology on Y

Having fixed a Spin^c structure $\mathfrak{s}$ on M, and a strongly $\mathfrak{s}$-admissible Heegaard diagram $(\Sigma,\mathbf{A},\mathbf{B},z)$ arising from a self-indexing Morse funcrion f and a gradient-like vector field $\mathfrak{v}$, we described in Section 2 a 3-manifold $Y \simeq M\#_G S^1\times S^2$, a stable Hamiltonian structure (a,w) with associated Reeb vector field $\mathcal{R}$, and an auxiliary 1-form $\hat{a}$ on Y. Next, we choose a Riemannian metric on Y so that

- $*\hat{a} = w$,
- $|\mathcal{R}| = 1$.

Choosing such a Riemannian metric on Y is equivalent to choosing a compatible complex structure on the 2-plane field $K^{-1} = ker(\hat{a})$ oriented by w. The latter determines a canonical Spin^c structure $\mathfrak{s}_0$ on Y satisfying

- $\langle c_1(\mathfrak{s}_0), [S^2]\rangle = 2$ where $[S^2]$ is the positive generator of $H_2(\mathcal{H}_0;\mathbb{Z})$,
- $\langle c_1(\mathfrak{s}_0), [S^2]\rangle = -2$ where $[S^2]$ is the positive generator of $H_2(\mathcal{H}_\mathfrak{p};\mathbb{Z})$ for some $\mathfrak{p} \in \Lambda$,
- $\mathfrak{s}_0 \otimes PD(\Gamma) = \underline{\mathfrak{s}}$.

The corresponding spinor bundle $\mathbb{S}_0$ splits into $\pm i$-eigenbundles of $\mathfrak{cl}(\hat{a})$ as $\underline{\mathbb{C}} \oplus K^{-1}$, where the left most summand is the $+i$-eigenbundle, and there exists a canonical (modulo action of $C^\infty(Y,S^1)$) Spin^c connection $\mathbb{A}_0$ on $\mathbb{S}_0$ such that $\mathcal{D}_{\mathbb{A}_0}(1_{\underline{\mathbb{C}}},0) = 0$. With the preceding understood, the spinor bundle $\underline{\mathbb{S}}$ associated to the Spin^c structure $\underline{\mathfrak{s}}$ splits into $\pm i$-eigenbundles of $\mathfrak{cl}(\hat{a})$ as $E \oplus (E \otimes K^{-1})$ where E is a Hermitian line bundle over Y with $c_1(E) = PD(\Gamma)$, and for any Spin^c connection $\mathbb{A}$ on $\underline{\mathbb{S}}$ we can write $\mathbb{A}^t = \mathbb{A}_0^t + 2A$ where A is a Hermitian connection on E. Now, we introduce a perturbed version of the Seiberg–Witten equations:

$$\begin{aligned} *F_A &= r(\psi^\dagger \tau \psi - i * w) - *\frac{1}{2}F_{\mathbb{A}_0}, \\ \mathcal{D}_A\psi &= 0, \end{aligned} \tag{18}$$

where $r > 0$ is a real number and ψ is a smooth section of $\underline{\mathbb{S}}$. If we write $\Psi = (2r)^{1/2}\psi$, the equations in (18) would become the same as the equations in (16) with $\varpi = -2rw$. Note that since $c_1(\underline{\mathfrak{s}}) = [w]$ in $H^2(Y;\mathbb{R})$, the above equations admit no reducible solutions for $r > \pi$; in fact, the equations are negatively monotone perturbed. In this case, a solution (A,ψ) of the equations in (18) is non-degenerate if the kernel of the following first order, elliptic, self-adjoint operator on $C^\infty(Y; iT^*Y \oplus \underline{\mathbb{S}} \oplus i\mathbb{R})$ has trivial kernel:

$$\mathcal{L}_{(A,\psi)}(b,\eta,\phi) = \begin{pmatrix} *db - d\phi - \left(\frac{r}{2}\right)^{1/2}(\psi^\dagger\tau\eta + \eta^\dagger\tau\psi) \\ \mathcal{D}_A\eta + (2r)^{1/2}(\mathfrak{cl}(b)\psi + \phi\psi) \\ *d*b - \left(\frac{r}{2}\right)^{1/2}(\eta^\dagger\psi - \psi^\dagger\eta) \end{pmatrix}. \tag{19}$$

One can achieve non-degeneracy of all irreducible solutions to the perturbed Seiberg–Witten equations in (18) via an additional exact perturbation with small norm from a certain Banach space Ω of smooth 1-forms. Having chosen $\mu \in \Omega$, the resulting equations would read

$$\begin{aligned} *F_A &= r(\psi^\dagger \tau \psi - i * w) - i * d\mu - *\frac{1}{2}F_{\mathbb{A}_0}, \\ \mathcal{D}_A\psi &= 0. \end{aligned} \tag{20}$$

Given a smooth section ψ of $\underline{\mathbb{S}}$, write $\psi = (\alpha,\beta)$. Recall that there exists a distinguished closed integral curve of $\mathcal{R}$ that intersects the Heegaard surface Σ at the base point z, denoted γ_z. Fix a Hermitian connection A_E on E that is flat on $\mathcal{H}_0$ and has holonomy 1 around γ_z, and a smooth non-decreasing function $\varphi : [0,\infty) \to [0,\infty)$ satisfying $\varphi(x) = 0$ for $0 \le x < \frac{7}{16}$ and $\varphi(x) = 1$ for $x \ge \frac{9}{16}$. Then, given a pair (A,ψ) of a Hermitian connection A on E and a smooth section ψ of $\underline{\mathbb{S}}$, define

$$\hat{A} := A - \frac{1}{2}\varphi(|\alpha|^2)|\alpha|^{-2}(\bar{\alpha}\nabla_A\alpha - \alpha\nabla_A\bar{\alpha}),$$
$$\mathrm{x}_z(A,\psi) := \int_{\gamma_z} \frac{i}{2\pi}(\hat{A} - A_E).$$

The function x_z is not invariant under the action of $C^\infty(Y,S^1)$, but it is invariant under the action of a subgroup $\mathcal{G}_{M_\Lambda}$ that is the kernel of the homomorphism $C^\infty(Y,S^1) \to \mathbb{Z}$ sending $u \in C^\infty(Y,S^1)$ to $\int_{\gamma_\tau} \frac{i}{2\pi}u^{-1}du$. A pair (A,ψ) of a Hermitian connection A on E and a smooth section of $\underline{\mathbb{S}}$ is called *holonomy non-degenerate* if $\mathrm{x}_z(A,\psi) - \frac{1}{2} \notin \mathbb{Z}$. There exists a residual set in Ω such that for every μ in this set and $r \in (\pi,\infty)$ in the complement of a discrete countable set dependent upon μ, all solutions to the equations in (20) are irreducible, non-degenerate, and holonomy non-degenerate. Moreover, the set $\mathcal{Z}_{SW,r}$ consisting of gauge equivalence classes of solutions to (20) is finite. With the preceding understood, the set $\hat{\mathcal{Z}}_{SW,r}$ consisting of $\mathcal{G}_{M_\Lambda}$-orbits of solutions to (20) admits a $1-1$ correspondence

$$\hat{\mathcal{Z}}_{SW,r} \leftrightarrow \mathcal{Z}_{SW,r} \times \mathbb{Z}, \tag{21}$$

where projection onto the $\mathbb{Z}$-factor is defined by the function x_z. Having said that, we define twisted versions of the Seiberg–Witten Floer cohomology on Y as follows: let $C^\infty_{SW,r}$ and $C^-_{SW,r}$ denote the free $\mathbb{Z}$-modules generated respectively by $\hat{\mathcal{Z}}_{SW,r}$ and the subset of $\hat{\mathcal{Z}}_{SW,r}$ identified with $\mathcal{Z}_{SW,r} \times \{\cdots,-2,-1\}$ via (21); and let $C^+_{SW,r}$ denote the quotient $C^\infty_{SW,r}/C^-_{SW,r}$. The relative grading on $C^\infty_{SW,r}$ is defined to be the negative of the spectral flow for a generic path of self-adjoint Fredholm operators starting and ending at the version of the operator in (19) for a solution of the equations in (20). Roughly speaking, the spectral flow of a generic path of self-adjoint Fredholm operators starting and ending at the associated versions of the operator in (19) is the net number of negative eigenvalues becoming positive.

The upcoming definition of the differential on the above defined Seiberg–Witten Floer chain complexes use solutions to the following flow equations:

$$\frac{\partial}{\partial s}A(s) = -*F_A(s) + r(\psi(s)^\dagger\tau\psi(s) - i*w) - *\frac{1}{2}F_{\mathbb{A}_0} - i*d\mu,$$
$$\frac{\partial}{\partial s}\psi(s) = -\mathcal{D}_{A(s)}\psi(s), \tag{22}$$

A solution $(A(s),\psi(s))$ of the above equations with $\lim_{s\to\pm\infty}(A(s),\psi(s)) = (A_\pm,\psi_\pm)$ being solutions to the equations in (20) is called an *instanton*. An instanton $(A(s),\psi(s))$ connecting two irreducible, non-degenerate solutions to the equations in (20) is called non-degenerate if the operator $\frac{\partial}{\partial s} + \mathcal{L}_{(A(s),\psi(s))}$ on $C^\infty(\mathbb{R}\times Y; iT^*Y \oplus \underline{\mathbb{S}} \oplus i\mathbb{R})$ has trivial cokernel. In order to make sure that all instantons are non-degenerate, we need to use certain generic perturbations from a Banach space $\mathcal{P}$ of *tame perturbations* as defined by Kronheimer and Mrowka in Section 11 of [17]. These are smooth gauge invariant functions on the configuration space with values in $iT^*M \oplus \mathbb{S}$. In particular, Ω is contained in $\mathcal{P}$. We suppress additional tame perturbations from the notation for sake of exposition. Given $r \in (\pi, \infty)$ and $\mu \in \Omega$ so that all solutions to the equations in (20) are irreducible and non-degenerate, we may choose tame perturbations to vanish to second order on any solution of the equations in (20) and at all points of a non-degenerate instanton with Fredholm index at most 2. We denote the subset of such perturbations in $\mathcal{P}$ by $\mathcal{P}_\mu$. There exists a residual subset of $\mathcal{P}_\mu$ such that all instantons are non-degenerate. For the remainder of this section, we assume that $\mu \in \Omega$, $r \in (\pi, \infty)$, and $\mathfrak{q} \in \mathcal{P}_\mu$ are chosen so that all solutions to the equations in (20) are irreducible, non-degenerate, holonomy non-degenerate, and all instanton solutions to the equations in (22) are non-degenerate. Having said that, a Fredholm index-0 instanton arises as a constant map from $\mathbb{R}$ to the moduli space of solutions to the equations in (20). The differential on $C^{\infty}_{SW,r}$ is defined by

$$\partial^{SW,r}_{\infty}[A_+, \psi_+] = \sum_{[A_-,\psi_-] \in \hat{\mathcal{Z}}_{SW,r}} \sigma_1([A_+, \psi_+], [A_-, \psi_-])[A_-, \psi_-],$$

where $\sigma_1([A_+, \psi_+], [A_-, \psi_-])$ is a signed count, modulo the action of the Lie group $\mathcal{G}_{M_\Lambda}$ and translations along the $\mathbb{R}$-factor, of Fredholm index-1 instantons $(A(s), \psi(s))$ with $\lim_{s\to+\infty}(A(s), \psi(s)) = (A_+, \psi_+)$ and $\lim_{s\to-\infty}(A(s), \psi(s)) = u \cdot (A_-, \psi_-)$ for some u in $\mathcal{G}_{M_\Lambda}$. There exists a certain closed 2-form on Y representing the Poincaré dual of $[\gamma_z]$ such that

$$\mathrm{x}_z(A_+, \psi_+) - \mathrm{x}_z(A_-, \psi_-) = \int_{\mathbb{R}\times Y} \frac{i}{2\pi} F_{\hat{A}(s)} \wedge \varpi_z \geq 0 \tag{23}$$

(see Proposition 7.1 in [21]). Therefore, the differential $\partial^{SW,r}_{\infty}$ preserves the filtration induced by the function x_z, and hence induces differentials on $C^{-}_{SW,r}$ and $C^{+}_{SW,r}$. The resulting homologies are denoted respectively by $H^{\infty}_{SW,r}$, $H^{-}_{SW,r}$, and $H^{+}_{SW,r}$. The latter fit into a long-exact sequence induced by the natural short-exact sequence

$$0 \longrightarrow C^{-}_{SW,r} \xrightarrow{\ i\ } C^{\infty}_{SW,r} \xrightarrow{\ \pi\ } C^{+}_{SW,r} \longrightarrow 0. \tag{24}$$

Meanwhile, a degree 2 chain map on $C^{\infty}_{SW,r}$ is defined by

$$U[A_+, \psi_+] = \sum_{[A_-,\psi_-] \in \hat{\mathcal{Z}}_{SW,r}} \sigma_{2,x}([A_+, \psi_+], [A_-, \psi_-])[A_-, \psi_-],$$

where $\sigma_{2,x}([A_+, \psi_+], [A_-, \psi_-])$ is a signed count, modulo the action of the Lie group $\mathcal{G}_{M_\Lambda}$ and translations along the $\mathbb{R}$-factor, of Fredholm index-2 instantons $(A(s), \psi(s))$ such that $\lim_{s\to+\infty}(A(s), \psi(s)) = (A_+, \psi_+)$ and $\lim_{s\to-\infty}(A(s), \psi(s)) = u \cdot (A_-, \psi_-)$ for some u in $\mathcal{G}_{M_\Lambda}$, and $\psi(0)(x) = (0, \beta(0)(x))$. Similarly, we can define degree 1 chain maps on $C^{\infty}_{SW,r}$ for any given element of $H_1(Y;\mathbb{Z})/torsion$ by a weighted sum of the algebraic intersection numbers of $\alpha(s)^{-1}(0)$ from Fredholm index-1 instantons $(A(s), \psi(s) = (\alpha(s), \beta(s)))$ with the cylinder over a cycle representing that element. As a result, $H^{\infty}_{SW,r}$, $H^{-}_{SW,r}$, and $H^{+}_{SW,r}$ each has the structure of a $\mathbb{Z}[U] \otimes \wedge^*(H_1(Y;\mathbb{Z})/torsion)$-module.

For any one of the flavors $H^{\infty}_{SW,r}$, $H^{-}_{SW,r}$, and $H^{+}_{SW,r}$, there are canonical isomorphisms between the homology groups corresponding to two different choices of data $(r_0, \mu_0, \mathfrak{q}_0)$

and $(r_1, \mu_1, \mathfrak{q}_1)$. In order to prove this, consider a generic path $\{(r_s, \mu_s, \mathfrak{q}_s)\}_{s\in[0,1]}$ of such data to be used with the equations in (20). Then, Kronheimer and Mrowka explain in Section 25 of [17] how to construct a chain map between the corresponding versions of the chain complexes by counting Fredholm index-0 instantons. These chain maps induce isomorphisms on homology that are independent of the choice of the generic path $\{(r_s, \mu_s, \mathfrak{q}_s)\}_{s\in[0,1]}$. Therefore, the homology groups corresponding to the perturbation data $(r_0, \mu_0, \mathfrak{q}_0)$ and $(r_1, \mu_1, \mathfrak{q}_1)$ are canonically isomorphic; in particular, the subscript r can be dropped from the notation. For the remainder of this article we denote the three flavors of the twisted version of Seiberg–Witten Floer cohomology of Y by H^{∞}_{SW}, H^{-}_{SW}, and H^{+}_{SW}, respectively.

With the preceding understood, the aim of this section is to explain what goes into the proof of the following theorem:

Theorem 4.1 (Theorem 3.4 in [18]). *There exists a commutative diagram*

$$\begin{array}{ccccccccc} \cdots & \longrightarrow & ech^{-} & \longrightarrow & ech^{\infty} & \longrightarrow & ech^{+} & \longrightarrow & \cdots \\ & & \downarrow & & \downarrow & & \downarrow & & \\ \cdots & \longrightarrow & H^{-}_{SW} & \longrightarrow & H^{\infty}_{SW} & \longrightarrow & H^{+}_{SW} & \longrightarrow & \cdots \end{array}$$

where the top row is the long-exact sequence for ech associated to (4) and the bottom row is the long-exact sequence for the twisted version of the Seiberg–Witten Floer cohomology of Y associated to (24) while the vertical arrows are canonical isomorphisms that reverse the sign of the relative gradings and intertwine the $\mathbb{Z}[U]\otimes\wedge^*(H_1(M;\mathbb{Z})/torsion)$-module structures.

The proof of the above theorem is along the same lines as Taubes's proof of the equivalence between ECH and Seiberg–Witten Floer cohomology [34, 35, 36, 37, 38], but in many ways it also borrows from the ideas involved in the proof of the equivalence between periodic Floer homology and Seiberg–Witten Floer cohomology by Lee and Taubes [24]. In particular, a priori estimates on the energy of instanton solutions to (22) follow from arguments that are exact analogs of those in the latter. The overarching idea is to filter the embedded contact homology chain complex via the symplectic action and define monomorphisms from each filtration level into the Seiberg–Witten Floer cochain complex. The desired isomorphism then follows from taking the direct limit. Next we explain how the aforementioned homomorphisms are constructed in the present setting. For precise statements of the results summarized in this section that lead to a proof of Theorem 4.1, see Theorem 1.5 in [21].

A filtration on the chain complex ecc^{∞} is defined by a $\mathbb{Z}$-valued function on $\hat{\mathcal{Z}}_{ech,M}$. For a given $[\Theta, Z] \in \hat{\mathcal{Z}}_{ech,M}$ where Θ is identified with $(v, \{(m_{\mathfrak{p}}, o_{\mathfrak{p}})\}_{\mathfrak{p}\in\Lambda})$ via (9), this function assigns to it the integer $\sum_{\mathfrak{p}\in\Lambda}|m_{\mathfrak{p}}| + 2|o_{\mathfrak{p}}|_{\mathrm{o}}$ where $|0|_{\mathrm{o}} = 0$, $|\pm 1|_{\mathrm{o}} = 1$, and $|\{1,-1\}|_{\mathrm{o}} = 2$. Given a finitely generated subgroup $\mathbb{H}$ of H^{∞}_{SW}, there exists $L_{\mathbb{H}} > 0$ such that for any $L \geq L_{\mathbb{H}}$ there exists $L' \geq L$ with the following significance: suppose $r > \pi$ is sufficiently large and $(\mu, \mathfrak{q}) \in \Omega\times\mathcal{P}$ is suitably generic subject to the constraints discussed in the previous subsection. Denote by $\hat{\mathcal{Z}}^{L'}_{ech,M}$ the subset of $\hat{\mathcal{Z}}_{ech,M}$ consisting of elements with filtration level less or equal L' and by $ecc^{\infty,L'}$ the free $\mathbb{Z}$-module that it generates. Then, there exists an injective $\mathbb{Z}$-equivariant map $\hat{\Phi}_r : \hat{\mathcal{Z}}^{L'}_{ech,M} \to \hat{\mathcal{Z}}_{SW,r}$ that induces a monomorphism

$$\mathbb{L}_r : ecc^{\infty,L'} \to C^{\infty}_{SW,r},$$

which intertwines the respective differentials and the $\mathbb{Z}[U]\otimes\wedge^*(H_1(Y;\mathbb{Z})/torsion)$-actions. Furthermore, the image of the induced map on homology contains $\mathbb{H}$. The assertion about the respective differentials and the $\mathbb{Z}[U]\otimes\wedge^*(H_1(Y;\mathbb{Z})/torsion)$-actions requires also the construction of an $\mathbb{R}$-equivariant map Ψ_r between the moduli spaces of J-holomorphic curves and instantons. The latter map together with the interpretation of the right-hand side of (23) as the algebraic intersection number of a J-holomorphic curve in $\mathbb{R}\times Y$ with the cylinder $\mathbb{R}\times\gamma_z$ verifies the claim about the remaining two equivalences and the long-exact sequences. The construction of both $\hat{\Phi}_r$ and Ψ_r make use of solutions to the vortex equations on $\mathbb{C}$. Next we talk about these equations and properties of their solutions.

4.3. Vortices on $\mathbb{C}$

Given a pair of a smooth imaginary-valued 1-form A_0 and a smooth complex-valued function α_0 on $\mathbb{C}$, the vortex equations on $\mathbb{C}$ read:

$$\begin{aligned} *dA_0 &= -i(1-|\alpha_0|^2), \\ \bar{\partial}_{A_0}\alpha_0 &= 0. \end{aligned} \tag{25}$$

Here, we assume that the complex plane is equipped with the standard Euclidean metric, and $\bar{\partial}_{A_0}$ denotes the ∂-bar operator on smooth sections of the trivial complex line bundle over $\mathbb{C}$. Recall that a smooth imaginary valued 1-form on $\mathbb{C}$ determines a holomorphic structure on the trivial complex line bundle over $\mathbb{C}$. These equations are subject to a pair of constraints:

- $|\alpha_0|^2 \le 1$,
- $\int_{\mathbb{C}}(1-|\alpha_0|^2)dvol = 2\pi m$ for some positive integer m.

A few quick observations are in order. First, if (A_0,α_0) is a solution of the equations in (25) subject to the constraint in the first bullet, then it follows from the strong maximum principle that either $|\alpha_0|<1$ or $|\alpha_0|\equiv 1$. Second, $|\alpha_0|$ has no non-zero local minimum unless $|\alpha_0|\equiv 1$. Assuming the contrary, in a neighborhood of a local minimum of $|\alpha_0|$ we can write $|\alpha_0| = e^u$ for a smooth function $u<0$. Then, the top equation in (25) implies that $-\Delta u = 1-e^{2u}$. At a non-zero local minimum of $|\alpha_0|$, the right-hand side of the latter equation is strictly positive, while the left-hand side is strictly negative. Therefore we get a contradiction. On the other hand, using the constraint in the second bullet above, we deduce that $1-|\alpha_0|^2 \le c_0 e^{-\sqrt{2}dist(\cdot,\alpha_0^{-1}(0))}$ where $c_0>1$ is a constant depending only on the integer m.

The group $C^\infty(\mathbb{C};S^1)$ acts on the space $C^\infty(\mathbb{C};iT^*\mathbb{C}\oplus\underline{\mathbb{C}})$ according to the following rule:

$$u\cdot(A_0,\alpha_0) = (A_0 - u^{-1}du, u\alpha_0).$$

Note that the set of solutions to the equations in (25) remain unchanged under this action. Meanwhile, the zeroes of α_0 are isolated and they appear as roots of the polynomial

$$z^m + \sigma_1 z^{m-1} + \cdots + \sigma_{m-1} z + \sigma_m,$$

where each σ_k is a function on $C^\infty(\mathbb{C}; iT^*\mathbb{C} \oplus \underline{\mathbb{C}})$ defined by

$$\sigma_k(A, \alpha_0) = \frac{1}{2\pi}\int_{\mathbb{C}} z^k(1 - |\alpha_0|^2)\, dvol$$

Denote by $\mathfrak{C}_m$ the space of $C^\infty(\mathbb{C}; S^1)$-orbits of solutions to the equations in (25) subject to the above constraints. Then $\mathfrak{C}_m$ is diffeomorphic to $\mathbb{C}^m$ via a map sending $[A_0, \alpha_0]$ to $(\sigma_1(A_0, \alpha_0), \ldots, \sigma_m(A_0, \alpha_0))$. The tangent space to $\mathfrak{C}_m$ at a point $[A_0, \alpha_0]$ is given by pairs of smooth complex-valued L^2 functions in the kernel of the following bounded complex-linear Fredholm operator on $C^\infty(\mathbb{C}; \mathbb{C} \oplus \mathbb{C})$,

$$\vartheta(q, \zeta) = (\partial q + 2^{-1/2}\bar{\alpha}_0\zeta, \bar{\partial}_{A_0}\zeta + 2^{-1/2}\alpha_0 q).$$

The tangent space admits a Hermitian inner product defined by

$$\langle (q, \zeta), (q', \zeta')\rangle = \frac{1}{\pi}\int_{\mathbb{C}} (\bar{q}q' + \bar{\zeta}\zeta')\, dvol.$$

which induces a Kähler metric on $\mathfrak{C}_m$. This metric coincides with the standard Kähler metric on $\mathbb{C}^m$ if $m = 1$. The formal L^2-adjoint of the operator ϑ is defined by

$$\vartheta^\dagger(p, \xi) = (-\bar{\partial}p + 2^{-1/2}\bar{\alpha}_0\xi, -\partial_{A_0}\xi + 2^{-1/2}\alpha_0 p).$$

Hence,

$$\vartheta\vartheta^\dagger(p, \xi) = (-\partial\bar{\partial}p + 2^{-1/2}|\alpha_0|^2\xi, -\bar{\partial}_{A_0}\partial_{A_0}\xi + 2^{-1/2}|\alpha_0|^2 p),$$

when (A_0, α_0) defines a solution of the equations in (25). For more details on properties of solutions to the vortex equations, the reader is encouraged to refer to [39] and [15].

Given $\mathrm{v} \in C^\infty(S^1; \mathbb{R})$ and $\mu \in C^\infty(S^1; \mathbb{C})$, we can define a Hamiltonian function $\mathfrak{h}$ on $\mathfrak{C}_m$ as follows:

$$\mathfrak{h}([A_0, \alpha_0]) = \frac{1}{4\pi}\int_{\mathbb{C}} (2\mathrm{v}|z|^2 + \mu\bar{z}^2 + \bar{\mu}z^2)(1 - |\alpha_0|^2)\, dvol.$$

The imaginary part of the Kähler metric on $\mathfrak{C}_m$ defines a symplectic form, therefore the function $\mathfrak{h}$ defines a Hamiltonian vector field on $\mathfrak{C}_m$ whose integral curves $\mathfrak{c}(t)$ obey

$$\frac{i}{2}\mathfrak{c}_*\left(\frac{d}{dt}\right)^{(1,0)} + \nabla^{(1,0)}\mathfrak{h}|_{\mathfrak{c}(t)} = 0, \tag{26}$$

where $\nabla^{(1,0)}$ denotes the holomorphic part of the gradient. A solution $\mathfrak{c}(t)$ of the equation in (26) satisfying $\mathfrak{c}(t + T) = \mathfrak{c}(t)$ for some $T > 0$ is called *periodic*. Given a smooth map $\mathfrak{c} : S^1 \to \mathfrak{C}_m$, the bundle $\mathfrak{c}^* T_{1,0}\mathfrak{C}_m$ admits a Hermitian connection arising from the Levi-Civita connection on $T_{1,0}\mathfrak{C}_m$. A periodic solution $\mathfrak{c}$ of the equation in (26) is called *non-degenerate* if the linearization of the Hamiltonian flow, namely the operator on $C^\infty(S^1; \mathfrak{c}^* T_{1,0}\mathfrak{C}_m)$ sending ζ to $\frac{i}{2}\nabla_{S^1}\zeta + (\nabla_\zeta \nabla^{(1,0)}\mathfrak{h})|_{\mathfrak{c}}$, has trivial kernel. Here ∇_{S^1} denotes the covariant derivative for the Hermitian connection on $\mathfrak{c}^* T_{1,0}\mathfrak{C}_m$.

Now let γ be a closed integral curve of the Reeb vector field $\mathcal{R}$, and fix a unitary trivialization of $K^{-1}|_\gamma$. Using the latter and the exponential map, identify a tubular neighborhood of γ in Y with a tubular neighborhood of $\gamma \times \{0\}$ in $\gamma \times \mathbb{C}$. Let $\varrho > 0$ be the radius of the latter, and (t, z) denote the coordinates on $\gamma \times \mathbb{C}$ where t is the affine coordinate on γ with values in $\mathbb{R}/L_\gamma\mathbb{Z}$, L_γ being the length of γ, and z is the complex coordinate on $\mathbb{C}$. We can arrange that the linearization of the aforementioned identification along $\gamma \times \{0\}$ sends $\frac{\partial}{\partial t}$ to $\frac{L_\gamma}{2\pi}\mathcal{R}$. In these coordinates, w can be written as

$$w = \frac{i}{2} dz \wedge d\bar{z} - 2(\nu z + \mu\bar{z})(d\bar{z} \wedge dt) - 2(\nu\bar{z} + \mu z)(dz \wedge dt) + \cdots,$$

where ν is a real-valued function on S^1, since w is a closed 2-form, μ is a complex-valued functions on S^1, and the norms of the unwritten terms are bounded by $c_0|z|^2$. With the preceding understood, the next lemma is crucial to the construction of the map $\hat{\Phi}_r$.

Lemma 4.2 (Lemma 3.2 in [21]). *Suppose that $\Theta \in \mathcal{Z}_{ech,M}$ and γ is a closed integral curve of $\mathcal{R}$ from Θ. Then for $m = 1$ there exists a unique periodic solution of the equation in (26) associated to the pair (ν, μ) arising from a tubular neighborhood of γ as above. This solution arises as the constant map from S^1 to $\mathfrak{C}_m$ represented by a pair (A_0, α_0) such that $\alpha_0^{-1}(0) = 0$, and it is non-degenerate. Moreover, for any $\mathfrak{p} \in \Lambda$ and $\gamma \in \{\gamma_\mathfrak{p}^+, \gamma_\mathfrak{p}^-\}$, there are no periodic solutions to the equations in (26) if $m > 1$.*

4.4. Construction of the map $\hat{\Phi}_r$

The map $\hat{\Phi}_r$ arises as the $\mathbb{Z}$-equivariant covering of a map Φ_r from $\mathcal{Z}^{L'}_{ech,M}$ to $\mathcal{Z}_{SW,r}$. The latter map is constructed in two steps. First, use periodic solutions to the equation in (26) for $m = 1$ in order to associate to each $\Theta \in \mathcal{Z}^{L'}_{ech,M}$ the gauge equivalence class of a pair (A_Θ, ψ_Θ) of a Hermitian connection on E and a smooth section of $\mathbb{S}$ that nearly solves the equations in (20). Second, use perturbation theory for linear operators to show existence and uniqueness, up to gauge, of an honest solution of the equations in (20) associated to Θ. Various properties of the map Φ_r are summarized in the following proposition.

Proposition 4.3 (Proposition 3.1 in [21]). *There exists $\kappa > \pi$ such that for any $\mathrm{E} > 1$ and $L > \kappa\mathrm{E}$, there exists $\kappa_L > \kappa$ satisfying the following: fix $r \geq \kappa_L$ and $\mu \in \Omega$ with small norm. Then there exists an injective map $\Phi_r : \mathcal{Z}^L_{ech,M} \to \mathcal{Z}_{SW,r}$ whose image contains gauge equivalence classes of $(A, \psi = (\alpha, \beta))$ with*

$$\mathrm{M}(A, \psi = (\alpha, \beta)) = r \int_Y (1 - |\alpha|^2) dvol < \mathrm{E}.$$

Moreover, if $(A, \psi = (\alpha, \beta))$ is a solution of the equations in (20) whose gauge equivalence class is $\Phi_r(\Theta)$ for some $\Theta \in \mathcal{Z}^L_{ech,M}$, then

- *$(A, \psi = (\alpha, \beta))$ is both non-degenerate and holonomy non-degenerate.*
- *$\mathrm{M}(A, \psi = (\alpha, \beta)) < 2\pi \sum_{\gamma \in \Theta} L_\gamma + \kappa^{-1}$.*

- *$\alpha^{-1}(0)$ is a disjoint union of embedded closed curves which are in $1-1$ correspondence with components of Θ, and each lies in a $\kappa r^{-1/2}$ radius tubular neighborhood of its partner from Θ.*
- *The map Φ_r extends to an injective map from $\mathcal{Z}^L_{ech,M} \times \mathbb{Z} \to \mathcal{Z}_{SW,r} \times \mathbb{Z}$ which is identity when restricted to the $\mathbb{Z}$-factors. The latter map defines a $\mathbb{Z}$-equivariant map $\hat{\Phi}_r : \hat{\mathcal{Z}}^L_{ech,M} \to \hat{\mathcal{Z}}_{SW,r}$ via the identification in (9) and the one defined by x_z.*

A priori estimates on solutions to the equations in (20)—such as those in the next lemma—are crucial to the proof of the above proposition.

Lemma 4.4 (Lemma 2.1 in [21]). *There exists $\kappa > 1$ such that if $r \geq \kappa$ and $\mu \in \Omega$ has small norm, then a solution $(A, \psi = (\alpha, \beta))$ of the equations in (20) satisfies the following:*

- $|\alpha| \leq 1 + \frac{\kappa}{r}$,
- $|\beta|^2 \leq \frac{\kappa}{r}(1 - |\alpha|^2) + \frac{\kappa^3}{r^2}$,
- $|\nabla^A \alpha|^2 \leq \kappa r(1 - |\alpha|^2) + \kappa^3$,
- $|\nabla^{\mathbb{A}_0^t + A}\beta|^2 \leq \kappa(1 - |\alpha|^2) + \frac{\kappa^3}{r}$.

Furthermore, for any $i \geq 1$, there exists $\kappa_i > 1$ such that if $r \geq \kappa_i$ and $\mu \in \Omega$ has small norm, then

- $|(\nabla^A)^i \alpha| + r^{1/2}|(\nabla^{\mathbb{A}_0^t + A})^i \beta| \leq \kappa_i r^{i/2}$.

The proof of the above lemma is an application of the Bochner–Weitzenböck formula for the Dirac operator and standard elliptic regularity arguments. The above estimates are exact analogs of those in Lemmas 2.2-2.4 of [33] and in Section 2e of [31]. An improvement over the above lemma's estimates is stated as follows:

Lemma 4.5 (Lemma 2.3 in [21]). *There exists $\kappa > 1$ with the following property: suppose $r \geq \kappa$ and $\mu \in \Omega$ has small norm. Let $(A, \psi = (\alpha, \beta))$ be a solution of the equations in (20), and $Y_* \subset Y$ denote the subset of points where $1 - |\alpha|^2 \geq \kappa^{-1}$. Then*

$$|1 - |\alpha|^2| \leq e^{-\frac{\sqrt{r}dist(\cdot, Y_*)}{\kappa}} + \kappa r^{-1}.$$

In addition to the above a priori estimates on a solution $(A, \psi = (\alpha, \beta))$ of the equations in (20), we deduce several facts about the set $\alpha^{-1}(0)$. What follows is part of the content of Proposition 2.4 in [21]. There exists $\kappa > \pi$ with the following property: suppose $r \geq \kappa$ and $\mu \in \Omega$ has small norm. Let $(A, \psi = (\alpha, \beta))$ be a solution of the equations in (20), and $Y_r \subset Y$ denote the subset of points with distance $\kappa^2 r^{-1/2}$ from the curves in $\cup_{\mathrm{p} \in \Lambda}\{\gamma_{\mathrm{p}}^+, \gamma_{\mathrm{p}}^-\}$. Then, α is transversal to the zero section of the Hermitian bundle E over the closure of Y_r. Moreover, $\alpha^{-1}(0)$ consists of at most G components each of which is either a properly embedded arc or a circle, and it has the following peoperties:

- The tangent line to each component has distance at most $\kappa r^{-1/2}$ from the line along the Reeb vector field $\mathcal{R}$.
- The intersection of each component with any $\mathcal{H}_{\mathrm{p}}$ lies where $1 - 3\cos^2\theta > 0$.

- The intersection of $\alpha^{-1}(0)$ with $M_\delta \subset Y_r$ consists of G properly embedded segments that connect, in a $1-1$ manner, the boundaries of the radius δ coordinate balls around index-1 and index-2 critical points of the Morse function f.

Next we briefly explain how the map Φ_r as in Proposition 4.3 is constructed, following the general philosophy due to Taubes. The first of the two steps in the construction is to use non-degenerate periodic solutions to the equation in (26) to construct a pair of a Hermitian connection on E and a smooth section of $\underline{\mathbb{S}}$ on a neighborhood of each component of Θ, and then extend it to the rest of Y in a canonical manner. This is made possible by Lemma 4.2. The resulting pair nearly solves the equations in (20). In order to explain how such a pair is constructed, we start with a preliminary digression. The point in $\mathfrak{C}_1$ identified with the origin in $\mathbb{C}$ via the identification described in the previous subsection is called the *symmetric vortex*, because it is invariant under the action of S^1 by rotations on $\mathbb{C}$. There exists a unique pair (A_0, α_0) which defines the symmetric vortex and can be written as

- $A_0 = A_{\underline{\mathbb{C}}} - \frac{1}{2}\mathrm{a}_0(z^{-1}dz - \bar{z}^{-1}d\bar{z})$,
- $\alpha_0 = |\alpha_0|\frac{z}{|z|}$,

where $A_{\underline{\mathbb{C}}}$ is the trivial flat connection on the trivial complex line bundle over $\mathbb{C}$, and a_0 is a real-valued function on $\mathbb{C}$. Meanwhile, the following equation has a unique smooth L^2 solution

$$\vartheta\vartheta^\dagger(p,\xi) = -(\frac{1}{\sqrt{2}}(1-|\alpha_0|^2), \partial_{A_0}\alpha_0),$$

which can be written as $(p,\xi) = (2^{1/2}z\alpha_0^{-1}\partial_{A_0}\alpha_0, -\bar{z}\bar{\alpha}_0^{-1}(1-|\alpha_0|^2)$. Now, given Θ in $\mathcal{Z}^L_{ech,M}$ and $\gamma \in \Theta$, define a function $r_\gamma : \mathbb{C} \to \mathbb{C}$ by $r_\gamma(z) = (\frac{L_\gamma}{2\pi}r)^{1/2}z$. Fix $\rho \in (0, \frac{1}{100}\varrho)$ and let U_γ and U'_γ denote respectively the radius 4ρ and radius ρ tubular neighborhoods of γ. We choose ρ in such a way that $U_\gamma \cap U_{\gamma'} = \emptyset$ for any distinct pair $\gamma, \gamma' \in \Theta$. Next, denote by $U_0 = Y \setminus \cup_{\gamma\in\Theta}U'_\gamma$, and let $\chi_\gamma = \chi(|z|/\rho - 1)$. Note that the restriction of the Hermitian bundle E to any $U \in \{U_0\} \cup \{U_\gamma\}_{\gamma\in\Theta}$ admits a Hermitian trivialization so that the trivializing unit length section of the bundle $E|_{U_0}$ is identified with the section $\frac{z}{|z|}$ of the bundle $E|_{U_\gamma}$ under the corresponding trivialization. With the preceding understood, write a Hermitian connection A on $E|_U$ as $A = A_{\underline{\mathbb{C}}} + \mathrm{a}_U$ where a_U is an imaginary-valued 1-form on U, and a smooth section of $\underline{\mathbb{S}}|_U$ as (α_U, β_U). We will define a pair (A_Θ, ψ_Θ) of a Hermitian connection on E and a smooth section of $\underline{\mathbb{S}}$ via their restrictions to each $U \in \{U_0\} \cup \{U_\gamma\}_{\gamma\in\Theta}$.

- If $U = U_0$, then $\mathrm{a}_{U_0} = 0$ and $\psi_0 = (\alpha_{U_0}, \beta_{U_0}) = (1, 0)$.
- If $U = U_\gamma$ for some $\gamma \in \Theta$, then

$$\mathrm{a}_U = \chi_\gamma\left[i2^{1/2}\mathrm{v}r_\gamma^* p\, dt - \frac{1}{2}r_\gamma^*(\mathrm{a}_0(z^{-1}dz - \bar{z}^{-1}d\bar{z}))\right] - (1-\chi_\gamma)u_r^{-1}du_r$$

and $(\alpha_U, \beta_U) = (\chi_\gamma r_\gamma^*\alpha_0 + (1-\chi_\gamma)u_r, \chi_\gamma i\upmu r^{-1/2}r_\gamma^*\xi)$ where $u_r = r_\gamma^*(z/|z|)$. Denote the corresponding pair of Hermitian connection on $E|_{U_\gamma}$ and smooth section of $\underline{\mathbb{S}}|_{U_\gamma}$ by (A_γ, ψ_γ).

Since the transition function between trivializations of $E|_U$ and $E|_{U_\gamma}$ for $\gamma \in \Theta$ is given by $z/|z|$ over $U \cap U_\gamma$, the collection $\{(A_0, \psi_0), \{(A_\gamma, \psi_\gamma)\}_{\gamma \in \Theta}\}$ defines the desired pair (A_Θ, ψ_Θ).

With (A_Θ, ψ_Θ) in hand, write a solution (A, ψ) of the equations in (20) as

$$(A, \psi) = (A_\Theta, \psi_\Theta) + ((2r)^{1/2} b, \eta),$$

where b is an imaginary-valued 1-form on Y and η is a smooth section of $\underline{\mathbb{S}}$. The pair (A, ψ) solves the equations on (20) if and only if $\mathfrak{b} = (b, \eta, \phi) \in C^\infty(Y; iT^*Y \oplus \underline{\mathbb{S}} \oplus i\mathbb{R})$ solves a certain linear first order elliptic differential equation. This equation can be written as

$$\mathcal{L}_{(A_\Theta, \psi_\Theta)} \mathfrak{b} + r^{1/2} \mathfrak{b} * \mathfrak{b} = \mathfrak{u},$$

where $\mathfrak{b} \mapsto \mathfrak{b} * \mathfrak{b}$ is a quadratic bundle map of $iT^*Y \oplus \underline{\mathbb{S}} \oplus i\mathbb{R}$ and $\mathfrak{u}$ has bounded norm. A careful investigation of the eigenvalue equation $\mathcal{L}_{(A_\Theta, \psi_\Theta)} \mathfrak{b} = \lambda \mathfrak{b}$ yield a solution of the above equations (See Appendix to [21] for details). The latter is used to define $\Phi_r(\Theta)$ for sufficiently large $r > \pi$.

We end this section with a few brief remarks as to the properties of the map Φ_r listed in Proposition 4.3. The second and the third bullets of Proposition 4.3 follow from the construction of the map Φ_r and a priori estimates from Lemmas 4.4 and 4.5. The assertion in the first bullet of Proposition 4.3 that the image of the map Φ_r is represented by pairs that are holonomy non-degenerate follows from Lemma 4.2 and the observation that $\hat{A}$ is flat with covariant constant section $\alpha/|\alpha|$ near γ_z. Therefore, $\mathrm{x}_z(A, \psi)$ is an integer since the connection A_E on E was chosen to have trivial holonomy around γ_z. As for the fourth bullet, fix a cycle $Z \in H_2(Y; \Theta, \Theta_0)$ so that it has zero algebraic intersection number with the curve γ_z. Then, the construction of the map Φ_r associates to the equivalence class $[\Theta, Z]$ the $\mathcal{G}_{M_\Lambda}$-orbit of a solution (A, ψ) of the equations in (20) with $\mathrm{x}_z(A, \psi) = 0$.

5. Back to the Seiberg–Witten Floer homology of M

The goal of this section is to explain the relation between the Seiberg–Witten Floer homology of M and the twisted Seiberg–Witten Floer cohomology of Y as described in Section 4.2. More precisely, we discuss the following filtered connected sum theorem for Seiberg–Witten Floer homology:

Theorem 5.1 (Theorem 4.1 in [18]). *There exists a commutative diagram*

$$\begin{array}{ccccccccc}
\cdots \to & H^-_{SW} & \to & H^\infty_{SW} & \to & H^+_{SW} & \to \cdots \\
& \downarrow & & \downarrow & & \downarrow & \\
\cdots \to & \widehat{HM}_*(M, \mathfrak{s}, c_b) \otimes_{\mathbb{Z}} \hat{\mathbb{V}}^{\otimes G} & \to & \overline{HM}_*(M, \mathfrak{s}, c_b) \otimes_{\mathbb{Z}} \hat{\mathbb{V}}^{\otimes G} & \to & \widetilde{HM}_*(M, \mathfrak{s}, c_b) \otimes_{\mathbb{Z}} \hat{\mathbb{V}}^{\otimes G} & \to \cdots
\end{array}$$

where the top row is the long-exact sequence for the twisted version of the Seiberg–Witten Floer cohomology of Y associated to (24) and the bottom row is the standard long-exact sequence for the Seiberg–Witten Floer homology of M with balanced perturbations tensored with $\hat{\mathbb{V}}^{\otimes G}$, while the vertical arrows are isomorphisms that reverse the sign of the relative gradings and intertwine the $\mathbb{Z}[U] \otimes \wedge^*(H_1(M; \mathbb{Z})/torsion)$-module structures.

The above theorem follows from an S^1-equivariant formulation of Seiberg–Witten Floer homology and a connected sum formula. Before we state these results, we need some notation and terminology. In this regard, let (C_*, ∂_*) be a chain complex of $\mathbb{Z}$-modules that admits a degree -2 chain map U. Then one may consider the mapping cone of the U-map defined by

$$(S_U(C_*), S_U(\partial_*)) := (C_* \otimes_{\mathbb{Z}} \mathbb{Z}[W]/(W^2), \partial_* \otimes \sigma + U \otimes W),$$

where $\sigma(p + qW) := p - qW$ and $\mathbb{Z}[W]/(W^2)$ is graded so that W is of degree 1. Hence, multiplication by $1 \otimes W$ defines a degree 1 chain map. Conversely, introduce the $\mathbb{Z}$-modules

$$\begin{aligned} \mathrm{V}^\infty &:= \mathbb{Z}[U, U^{-1}], \\ \mathrm{V}^- &:= U\mathbb{Z}[U], \\ \mathrm{V}^+ &:= \mathbb{Z}[U, U^{-1}]/U\mathbb{Z}[U]. \end{aligned}$$

Suppose (C_*, ∂_*) is a chain complex of $\mathbb{Z}$-modules that admits a degree 1 chain map W with $W^2 = 0$. Then we can define the following chain complexes:

$$\begin{aligned} E^\infty(C_*) &:= C_* \otimes_{\mathbb{Z}} \mathrm{V}^\infty, \\ E^-(C_*) &:= C_* \otimes_{\mathbb{Z}} \mathrm{V}^-, \\ E^+(C_*) &:= C_* \otimes_{\mathbb{Z}} \mathrm{V}^+, \end{aligned}$$

with the differential $E(\partial_*) := (\partial_* \otimes 1 + W \otimes U)$ and the U-map defined as multiplication by $1 \otimes U$ which is of degree -2. Note that there exists a short-exact sequence

$$0 \to \mathrm{V}^- \xrightarrow{i} \mathrm{V}^\infty \xrightarrow{\pi} \mathrm{V}^+ \to 0, \tag{27}$$

which induces a short-exact sequence

$$0 \to E^-(C_*) \xrightarrow{id \otimes i} E^\infty(C_*) \xrightarrow{id \otimes \pi} E^+(C_*) \to 0,$$

with associated long-exact sequence of homologies

$$\cdots \to H_*(E^-(C_*)) \to H_*(E^\infty(C_*)) \to H_*(E^+(C_*)) \to H_*(E^-(C)) \to \cdots. \tag{28}$$

Now, the next result is a kind of Kozsul duality in the sense of [10].

Proposition 5.2 (Proposition 4.8 in [22]). *Let (C_*, ∂_*) be a chain complex of $\mathbb{Z}$-modules that admits a degree -2 chain map U. Then there are isomorphisms*

$$H_*(E^\circ S_U(C_*)) \cong H_*(C_* \otimes_{\mathbb{Z}[U]} \mathrm{V}^\circ),$$

where $\circ \in \{\infty, -, +\}$. Furthermore, these isomorphisms are natural with respect to the associated long-exact sequences. On the other hand, let (C_*, ∂_*) be a chain complex of $\mathbb{Z}$-modules that admits a degree 1 chain map Y. Then there is an isomorphism

$$H_*(S_U E^-(C_*)) \cong H_*(C_*).$$

With the preceding understood, the first main ingredient in the proof of Theorem 5.1 is the following S^1-equivariant formulation of Seiberg–Witten Floer homologies due to Lee (cf. [23]):

Proposition 5.3 (Proposition 5.9 in [22]). *Let (C_*, ∂_*) denote the Seiberg–Witten Floer chain complex $(\widehat{CM}_*(M, \mathfrak{s}, c_b), \hat{\partial})$ where c_b indicates the use of a balanced perturbation. Then there exists a commutative diagram*

$$\begin{array}{ccccccc} \cdots \longrightarrow & H_*(E^- S_U(C_*)) & \longrightarrow & H_*(E^\infty S_U(C_*)) & \longrightarrow & H_*(E^+ S_U(C_*)) & \longrightarrow \cdots \\ & \downarrow & & \downarrow & & \downarrow & \\ \cdots \longrightarrow & \widehat{HM}_*(M, \mathfrak{s}, c_b) & \longrightarrow & \overline{HM}_*(M, \mathfrak{s}, c_b) & \longrightarrow & \widecheck{HM}_*(M, \mathfrak{s}, c_b) & \longrightarrow \cdots \end{array}$$

where the top row is the long-exact sequence for equivariant homologies as in (28) and the bottom row is the long-exact sequence for the Seiberg–Witten Floer homology of M with balanced perturbations, while the vertical arrows are isomorphisms that preserve the relative gradings and intertwine the $\mathbb{Z}[U] \otimes \wedge^*(H_1(M;\mathbb{Z})/torsion)$-module structures.

Remark 5.4. Kronheimer and Mrowka prove that the chain complexes $(\widehat{CM}_*(M, \mathfrak{s}, c_b), \hat{\partial})$ and $(\widehat{CM}_*(M, \mathfrak{s}, c_-), \hat{\partial})$, where c_- denotes the period class for some negatively monotone perturbation, are chain homotopy equivalent (see Theorem 31.5.1 in [17]). Therefore, in the above proposition, one can also take C_* to be $(\widehat{CM}_*(M, \mathfrak{s}, c_-), \hat{\partial})$.

Next we discuss the second main ingredient in the proof of Theorem 5.1, which is a connected sum formula for Seiberg–Witten Floer homology. Before we state the connected sum formula, we need to introduce further notation. Let M_1 and M_2 be two closed, connected, and oriented 3-manifolds. Fix Spin^c structures $\mathfrak{s}_1$ and $\mathfrak{s}_2$ on M_1 and M_2 respectively. Denote by $\mathfrak{s}_\#$ and $\mathfrak{s}_\sqcup$ respectively the unique Spin^c structures on $M_1 \# M_2$ and $M_1 \sqcup M_2$ induced by $\mathfrak{s}_1$ and $\mathfrak{s}_2$. Given $\Omega_1 \in H^2(M_1; \mathbb{R})$ and $\Omega_2 \in H^2(M_2; \mathbb{R})$ the cohomology classes for perturbations of the Seiberg–Witten equations on M_1 and M_2, denote by $\Omega_\#$ and $\Omega_\sqcup$ the corresponding cohomology classes in $H^2(M_1 \# M_2; \mathbb{R})$ and $H^2(M_1 \sqcup M_2; \mathbb{R})$ respectively.

Proposition 5.5 (Proposition 6.2 in [22]). *Suppose that $\Omega_\#$ yields a negatively monotone perturbation. Then the chain complexes*

$$\widehat{CM}_*(M_1 \# M_2, \mathfrak{s}_\#, c_{\Omega_\#}) \qquad \text{and} \qquad S_{U_\sqcup}(\widehat{CM}_*(M_1 \sqcup M_2, \mathfrak{s}_\sqcup, c_{\Omega_\sqcup}))$$

are chain homotopy equivalent. Furthermore, the induced isomorphism on homology intertwines the $\mathbb{Z}[U] \otimes \wedge^*(H_1(M_1;\mathbb{Z})/torsion)) \otimes \wedge^*(H_1(M_2;\mathbb{Z})/torsion)$-module structures and preserves the relative gradings.

Kronheimer and Mrowka defined Seiberg–Witten Floer chain complexes for connected 3-manifolds. In the above proposition and in what follows, we abuse the notation to denote by $(\widehat{CM}_*(M_1 \sqcup M_2, \mathfrak{s}_\sqcup, c_{\Omega_\sqcup}), \hat{\partial}_\sqcup)$ the tensor product of $(\widehat{CM}_*(M_1, \mathfrak{s}_1, c_{\Omega_1}), \hat{\partial}_{M_1})$ and $(\widehat{CM}_*(M_2, \mathfrak{s}_2, c_{\Omega_2}), \hat{\partial}_{M_2})$, with $U_\sqcup = U_1 \otimes 1 - 1 \otimes U_2$. This is acceptable since at least one of Ω_1 and Ω_2 yields a negatively monotone perturbation.

Now, Theorem 5.1 follows from an application of Proposition 5.5 to the following two cases:

- Take $M_1 = S^1 \times S^2$, $\mathfrak{s}_1$ so that $\langle c_1(\mathfrak{s}_1), [S^2] \rangle = 2$, and w_1 a nowhere zero closed 2-form representing $c_1(\mathfrak{s}_1)$ that is harmonic with respect to the product Riemannian metric on $S^1 \times S^2$. Meanwhile, take $M_2 = M$, $\mathfrak{s}_2$ any Spin^c structure, and w_2 a closed 2-form with non-degenerate zeros and $c_1(\mathfrak{s}_2) = [w_2]$ that is harmonic with respect to a suitable Riemannian metric on M, and has non-degenerate zeros in the case $c_1(\mathfrak{s}_2)$ is non-torsion. Denote by Y_0 the connected sum $M \# S^1 \times S^2$.
- Take $M_1 = Y_i = M \#_{i+1} S^1 \times S^2$ for $i \in \{0, 1, \ldots, \text{G} - 1\}$, $\mathfrak{s}_1$ a Spin^c structure with $c_1(\mathfrak{s}_1)$ non-torsion, and w_1 a closed 2-form representing $c_1(\mathfrak{s}_1)$ that is harmonic with respect to a suitable Riemannian metric on Y_i. Meanwhile, take $M_2 = S^1 \times S^2$, $\mathfrak{s}_2$ is the trivial Spin^c structure, and $w_2 \equiv 0$.

In either of the above cases, take $\Omega_1 = rc_1(\mathfrak{s}_1)$, $\Omega_2 = rc_1(\mathfrak{s}_2)$, and hence $\Omega_\# = rc_1(\mathfrak{s}_\#)$ for $r > \pi$ sufficiently large. Note that in each Y_i for $i \in \{0, 1, \ldots, \text{G} - 1\}$ and in $Y = Y_{\text{G}}$, there exists a special simple closed curve γ that represents the generator of $H_1(S^1 \times S^2; \mathbb{Z})$ for the $S^1 \times S^2$ summand with non-trivial Spin^c structure. This cycle introduces a local system of coefficients Π_γ (see Section 22.6 in [22]). With the preceding understood, there is also a filtered variant of Proposition 5.5:

Proposition 5.6 (Proposition 6.2 in [22]). *Let M_1 and M_2 be as in either of the above two cases. Then for each $\circ \in \{\infty, -, +\}$ there exists an isomorphism between*

$$H_*(\widehat{CM}_*(M_1 \# M_2, \mathfrak{s}_\#, c_{\Omega_\#}; \Pi_\gamma \otimes_{\mathbb{Z}} \mathrm{V}^\circ) \cong H_*(S_{U_\sqcup}(\widehat{CM}_*(M_1 \sqcup M_2, \mathfrak{s}_\sqcup, c_{\Omega_\sqcup}; \Pi_\gamma \otimes_{\mathbb{Z}} \mathrm{V}^\circ)).$$

Furthermore, these isomorphisms preserve the relative gradings, they intertwine the $\mathbb{Z}[U] \otimes \wedge^*(H_1(M_1; \mathbb{Z})/torsion)) \otimes \wedge^*(H_1(M_2; \mathbb{Z})/torsion)$-module structures, and are natural with respect to the long-exact sequences induced by (27).

In the case of the first bullet above, the equations in (20) on M_1 have a unique solution up to gauge. This solution can be written as $(A, (\alpha, \beta)) = (0, ((2r^{-1/2}, 0))$. Thus, for each $\circ \in \{\infty, -, +\}$ we have $\widehat{CM}_*(M_1, \mathfrak{s}_1, c_{\Omega_1}; \Pi_\gamma \otimes \mathrm{V}^\circ)) \cong \mathrm{V}^\circ$. Then the chain complex

$$S_{U_\sqcup}(\widehat{CM}_*(S^1 \times S^2 \sqcup M, \mathfrak{s}_\sqcup, c_{\Omega_\sqcup}; \Pi_\gamma \otimes_{\mathbb{Z}} \mathrm{V}^\circ)) = \mathrm{V}^\circ \otimes_{\mathbb{Z}} \widehat{CM}_*(M, \mathfrak{s}_2, c_{\Omega_2}) \otimes_{\mathbb{Z}} \mathbb{Z}[W]/(W^2),$$

with differential $S_{U_\sqcup}(\hat{\partial}_\sqcup) = 1 \otimes \hat{\partial}_M \otimes \sigma + (U_1 \otimes 1 - 1 \otimes U_2) \otimes W$, can be rewritten as

$$E^\circ(\widehat{CM}_*(M, \mathfrak{s}_2, c_{\Omega_2}) \otimes_{\mathbb{Z}} \mathbb{Z}[W]/(W^2), \hat{\partial}_M \otimes \sigma - U_2 \otimes W),$$

in other words, $E^\circ S_{U_2}(\widehat{CM}_*(M, \mathfrak{s}_2, c_{\Omega_2}), \hat{\partial}_M)$. Using Propositions 5.3 and 5.6, we obtain isomorphisms

$$\begin{aligned} H_*(\widehat{CM}_*(Y_0, \mathfrak{s}_\#, c_{\Omega_\#}; \Pi_\gamma \otimes_{\mathbb{Z}} \mathrm{V}^\infty) &\cong \overline{HM}_*(M, \mathfrak{s}, c_b), \\ H_*(\widehat{CM}_*(Y_0, \mathfrak{s}_\#, c_{\Omega_\#}; \Pi_\gamma \otimes_{\mathbb{Z}} \mathrm{V}^-) &\cong \widehat{HM}_*(M, \mathfrak{s}, c_b), \\ H_*(\widehat{CM}_*(Y_0, \mathfrak{s}_\#, c_{\Omega_\#}; \Pi_\gamma \otimes_{\mathbb{Z}} \mathrm{V}^+) &\cong \widecheck{HM}_*(M, \mathfrak{s}, c_b), \end{aligned}$$

which respect the associated long-exact sequences. To be more explicit, the long-exact sequence for the groups on the left arises from the short-exact sequence in (27).

With regard to the second bullet above, the Seiberg–Witten equations for the trivial Spin^c structure on $S^1 \times S^2$ equipped with a constant curvature product Riemannian metric have only reducible solutions, and the set of solutions modulo gauge is identified with the circle $H^1(S^1 \times S^2; i\mathbb{R})/2\pi i H^1(S^1 \times S^2; \mathbb{Z})$. Then, as is explained in Section 36 of [17], $\widehat{CM}_*(S^1 \times S^2, \mathfrak{s}_2) \cong \mathbb{Z}[U_2] \oplus \mathbb{Z}[U_2][1]$ as graded $\mathbb{Z}$-modules. Therefore, the chain complex $(S_{U_\sqcup}(\widehat{CM}_*(Y_i \sqcup S^1 \times S^2, \mathfrak{s}_\sqcup, c_{\Omega_\sqcup}; \Pi_\gamma \otimes \mathrm{V}^\circ)), S_{U_\sqcup}(\hat{\partial}_\sqcup))$ can be written as

$$\widehat{CM}_*(Y_i, \mathfrak{s}_1, c_{\Omega_1}; \Pi_\gamma \otimes_\mathbb{Z} \mathrm{V}^\circ)) \otimes_\mathbb{Z} (\mathbb{Z}[U_2] \oplus \mathbb{Z}[U_2][1]) \otimes_\mathbb{Z} \mathbb{Z}[W]/(W^2),$$

with differential

$$\hat{\partial}_{Y_i} \otimes 1 \otimes \sigma - (U_1 \otimes U_2) \otimes W.$$

A straightforward computation shows that

$$H_*(S_{U_\sqcup}(\widehat{CM}_*(Y_i \sqcup S^1 \times S^2, \mathfrak{s}_\sqcup, c_{\Omega_\sqcup}; \Pi_\gamma \otimes_\mathbb{Z} \mathrm{V}^\circ),)) \cong H_*(\widehat{CM}_*(Y_i, \mathfrak{s}_1, c_{\Omega_1}; \Pi_\gamma \otimes_\mathbb{Z} \mathrm{V}^\circ) \otimes_\mathbb{Z} \hat{\mathrm{V}}.$$

In conclusion,

$$\begin{aligned} H_*(\widehat{CM}_*(Y_\mathrm{G}, \mathfrak{s}_\#, c_{\Omega_\#}; \Pi_\gamma \otimes_\mathbb{Z} \mathrm{V}^\infty) &\cong \overline{HM}_*(M, \mathfrak{s}, c_b) \otimes_\mathbb{Z} \hat{\mathrm{V}}^{\otimes \mathrm{G}}, \\ H_*(\widehat{CM}_*(Y_\mathrm{G}, \mathfrak{s}_\#, c_{\Omega_\#}; \Pi_\gamma \otimes_\mathbb{Z} \mathrm{V}^-) &\cong \widehat{HM}_*(M, \mathfrak{s}, c_b) \otimes_\mathbb{Z} \hat{\mathrm{V}}^{\otimes \mathrm{G}}, \\ H_*(\widehat{CM}_*(Y_\mathrm{G}, \mathfrak{s}_\#, c_{\Omega_\#}; \Pi_\gamma \otimes_\mathbb{Z} \mathrm{V}^+) &\cong \widecheck{HM}_*(M, \mathfrak{s}, c_b) \otimes_\mathbb{Z} \hat{\mathrm{V}}^{\otimes \mathrm{G}}. \end{aligned}$$

Here, Y_G is diffeomorphic to the manifold Y, but with opposite orientation. It is equipped with a closed 2-form and a Riemannian metric as induced by the initial data in the second bullet above. Keeping in mind that there are canonical isomorphisms between the Seiberg–Witten Floer cohomology groups described here on Y and the corresponding filtered Seiberg–Witten Floer homology groups on $\overline{Y}$, the orientation reversed copy of Y, we have $H_*(\widehat{CM}_*(Y_\mathrm{G}, \mathfrak{s}_\#, c_{\Omega_\#}; \Pi_\gamma \otimes_\mathbb{Z} \mathrm{V}^\circ) \cong H^\circ_{SW}$, which follows by analyzing the Seiberg–Witten equations on the product cobordism $\mathbb{R} \times \overline{Y}$ equipped with a suitable self-dual 2-form and a Riemannian metric interpolating between the data on Y_G and the stable Hamiltonian structure data on $\overline{Y}$. This last equivalence finishes our outline of the proof of Theorem 5.1.

Remark 5.7. The orientation of the manifold Y fixed in this article and in [19, 20, 21] is the opposite of the orientation used in [18] and in [22]. To be more precise, the convention used in the latter is that the inclusion of M_δ into Y is an orientation preserving map.

6. Final Remarks

As is mentioned in the Introduction, both Heegaard Floer and Seiberg–Witten Floer homologies fit in the general framework of topological quantum field theories, but for a few important caveats. Having fixed a commutative ring with unity $\mathfrak{R}$, a TQFT $\mathcal{F}$ in dimension 3 associates to every closed, oriented, and smooth 3-manifold M a module $\mathcal{F}(M)$ over $\mathfrak{R}$, and to every compact, oriented, and smooth 4-manifold with boundary M it associates an element $\mathcal{F}(Z) \in \mathcal{F}(M)$. In particular, $\mathcal{F}$ associates the ground ring

$\mathfrak{R}$ to the empty 3-manifold. According to Atiyah [1], $\mathcal{F}$ should also satisfy the following axioms:

(1) It is functorial with respect to orientation preserving diffeomorphisms of 3- and 4-manifolds.
(2) It is involutary, in other words, the module associated to $-M$ is the dual of the one associated to M.
(3) It is multiplicative.

Let us explain the multiplicative axiom. Suppose that Z is a cobordism from M_1 to M_2, i.e., $\partial Z = -M_1 \sqcup M_2$. In this case, the multiplicative axiom requires that

$$\mathcal{F}(Z) \in \mathcal{F}(-M_1) \otimes_{\mathfrak{R}} \mathcal{F}(M_2).$$

Using the involutary axiom, the latter can be rewritten as $\mathcal{F}(Z) \in Hom_{\mathfrak{R}}(\mathcal{F}(M_1), \mathcal{F}(M_2))$. The multiplicative axiom also requires that cobordism maps are transitive under the composition of cobordisms. Note that if Z is a cobordism from $\emptyset$ to M, then $\mathcal{F}(Z)$ is in $Hom_{\mathfrak{R}}(\mathfrak{R}, \mathcal{F}(M)) \cong \mathcal{F}(M)$; and if Z is closed, then $\mathcal{F}(Z) \in \mathfrak{R}$. Alternatively, given a closed, oriented, and smooth 4-manifold X, one cuts X open along a closed, oriented, and smooth 3-manifold M so that it becomes the composition of two cobordisms $X = Z_1 \cup_M Z_2$ where Z_1 is a cobordism from $\emptyset$ to M and Z_2 is a cobordism from M to $\emptyset$. Then, $\mathcal{F}(X) \in \mathfrak{R}$ is obtained via the pairing $\langle \mathcal{F}(Z_1), \mathcal{F}(-Z_2) \rangle$.

Here are a few points where Heegaard Floer and Seiberg–Witten Floer homologies differ from the classical picture of a TQFT explained above. First of all, the homology groups and the homomorphisms between them are defined only for connected 3-manifolds and connected cobordisms between them. Second, neither the Heegaard Floer homology groups $HF^\infty(M)$, $HF^-(M)$, $HF^+(M)$ nor the Seiberg–Witten Floer homology groups $\overline{HM}_*(M)$, $\widehat{HM}_*(M)$, $\widecheck{HM}_*(M)$ as defined by the unperturbed Seiberg–Witten equations in (14) are functorial since an infinite number of Spin^c structures on a given cobordism may contribute non-trivially to its induced homomorphism. Hence, one should instead consider the completions $\mathbf{HF}^\infty(M)$, $\mathbf{HF}^-(M)$, $\mathbf{HF}^+(M)$ and $\overline{HM}_\bullet(M)$, $\widehat{HM}_\bullet(M)$, $\widecheck{HM}_\bullet(M)$[4] of these groups with respect to the filtrations induced by their $\mathbb{Z}[U]$-module structures. Even so, Heegaard Floer homology is not natural in any obvious way, i.e., it assigns to a closed, connected, and oriented 3-manifold a trio of groups up to isomorphism. This issue has recently been addressed by Juhász and Thurston in [16]. Granted naturality, the invariants for closed, connected, oriented, and smooth 4-manifolds with $b_2^+ > 0$ obtained via the above recipe using any of these groups are trivial. Therefore, a slightly different recipe was employed by Ozsváth and Szabó to define the closed 4-manifold invariants.

[4]Having fixed a Spin^c structure $\mathfrak{s}$ on M, Kronheimer and Mrowka proved that there are isomorphisms

$$\begin{aligned} \overline{HM}_\bullet(M,\mathfrak{s}) &\cong H_*(\overline{CM}_\bullet(M,\mathfrak{s},c_b),\bar{\partial}) \\ \widehat{HM}_\bullet(M,\mathfrak{s}) &\cong H_*(\widehat{CM}_\bullet(M,\mathfrak{s},c_b),\hat{\partial}) \\ \widecheck{HM}_\bullet(M,\mathfrak{s}) &\cong H_*(\widecheck{CM}_\bullet(M,\mathfrak{s},c_b),\check{\partial}), \end{aligned}$$

(see Theorem 31.1.1 in [17]).

Given a closed, connected, and oriented 3-manifold M and a Spin^c structure $\mathfrak{s}$ on M, recall the long-exact sequence for the Heegaard Floer homology groups:

$$\cdots \to HF^-(M,\mathfrak{s}) \xrightarrow{i_*} HF^\infty(M,\mathfrak{s}) \xrightarrow{\pi_*} HF^+(M,\mathfrak{s}) \xrightarrow{\delta} HF^-(M,\mathfrak{s}) \to \cdots .$$

The above long-exact sequence is induced by a short-exact sequence of chain complexes, and δ is the connecting homomorphism. The image of δ, equivalently, the kernel of i_*, is a finite rank subgroup of $HF^-(M,\mathfrak{s})$ denoted by $HF^-_{red}(M,\mathfrak{s})$. The latter is isomorphic, by exactness, to the cokernel of π_*, which is denoted by $HF^+_{red}(M,\mathfrak{s})$. If Z is a cobordism from M_1 to M_2 with $b_2^+ > 0$, then the map $HF^-(Z,\mathfrak{t}) : HF^-(M_1,\mathfrak{s}_1) \to HF^-(M_2,\mathfrak{s}_2)$ factors through $HF^-_{red}(M_2,\mathfrak{s}_2)$, while the map $HF^+(Z,\mathfrak{t}) : HF^+(M_1,\mathfrak{s}_1) \to HF^+(M_2,\mathfrak{s}_2)$ factors through $HF^+_{red}(M_1,\mathfrak{s}_1)$. With the preceding understood, let X be a closed, oriented, and smooth 4-manifold with $b_2^+ > 1$ and $\mathfrak{s}$ be a Spin^c structure on X. Then, Ozsváth and Szabó define the 4-dimensional invariant $\Phi_{X,\mathfrak{s}}$ as follows: fix a closed, connected, oriented, and smooth 3-manifold M embedded in X that separates X into two pieces each with $b_2^+ > 0$, and such that $H_2(M;\mathbb{Z})$ has trivial image in $H_2(X;\mathbb{Z})$. Upon excising an open ball in each of these pieces, we obtain the cobordisms Z_1 from S^3 to M and Z_2 from M to S^3. Finally, define $\Phi_{X,\mathfrak{s}}$ as the composition of the maps

$$\begin{aligned} HF^-(Z_1,\mathfrak{s}|Z_1) &: HF^-(S^3) \to HF^-_{red}(M,\mathfrak{s}|M), \\ HF^+(Z_2,\mathfrak{s}|Z_2) &: HF^+_{red}(M,\mathfrak{s}|M) \to HF^+(S^3), \end{aligned}$$

while identifying $HF^-_{red}(M,\mathfrak{s}|M)$ with $HF^+_{red}(M,\mathfrak{s}|M)$ via the connecting homomorphism δ. An analog of the preceding construction, perhaps using a local system of coefficients, can be repeated using the Seiberg–Witten Floer homology groups to capture the Seiberg–Witten invariants of X, which are defined by a signed count of solutions of the Seiberg–Witten equations on X (cf. Section 3.8 in [17]). As a matter of fact, Ozsváth and Szabó conjectured the following:

Conjecture 6.1 ([28]). *Let X be a closed, connected, oriented, and smooth 4-manifold with $b_2^+ > 1$ and $\mathfrak{s}$ be a Spin^c structure on X. Then, the invariant $\Phi_{X,\mathfrak{s}}$ agrees with the Seiberg–Witten invariant for X in the Spin^c structure $\mathfrak{s}$.*

Granted the above claim about the Seiberg–Witten invariants of closed, connected, oriented, and smooth 4-manifolds, and the fact that Heegaard Floer and Seiberg–Witten Floer homologies are isomorphic, what remains to be shown to prove Conjecture 6.1 is the *naturality* of these isomorphisms, i.e., that the isomorphisms between Heegaard Floer and Seiberg–Witten Floer homology groups intertwine respective cobordism maps. In order to prove naturality, one would like to define cobordism maps for *ech* and compare these to cobordism maps for Heegaard Floer and Seiberg–Witten Floer homologies. In order to do so, one needs to understand the component of the moduli space of pseudo-holomorphic curves in cobordisms with *ech* index 0. In general, as explained by Hutchings, the latter may have a very complicated structure as there may be *ech* index-0 broken curves with negative *ech* index levels. Luckily, such complications do not arise in the context described in this article thanks to the fact that all Reeb orbits in our setting are hyperbolic and the only homologically trivial Reeb orbits are positive hyperbolic. Therefore, contrary to the case of embedded contact homology, it should be possible to give a definition of cobordism maps for *ech* solely via pseudo-holomorphic curves. This is part of the content of work in progress by the author.

Acknowledgements: The author would like to thank the anonymous referee for numerous comments that helped improve the exposition.

References

1. Michael Atiyah, *Topological quantum field theories*, Inst. Hautes Études Sci. Publ. Math. (1988), no. 68, 175–186 (1989).
2. Vincent Colin, Paolo Ghiggini, and Ko Honda, *Equivalence of Heegaard Floer homology and embedded contact homology via open book decompositions*, Proc. Natl. Acad. Sci. USA **108** (2011), no. 20, 8100–8105.
3. ———, *The equivalence of Heegaard Floer homology and embedded contact homology via open book decompositions I*, preprint (2012), arXiv:1208.1074.
4. ———, *The equivalence of Heegaard Floer homology and embedded contact homology via open book decompositions II*, preprint (2012), arXiv:1208.1077.
5. ———, *The equivalence of Heegaard Floer homology and embedded contact homology III: from hat to plus*, preprint (2012), arXiv:1208.1526.
6. Simon K. Donaldson, *An application of gauge theory to four-dimensional topology*, J. Differential Geom. **18** (1983), no. 2, 279–315.
7. Simon K. Donaldson, *Polynomial invariants for smooth four-manifolds*, Topology **29** (1990), no. 3, 257–315.
8. Andreas Floer, *An instanton-invariant for 3-manifolds*, Comm. Math. Phys. **118** (1988), no. 2, 215–240.
9. ———, *Morse theory for Lagrangian intersections*, J. Differential Geom. **28** (1988), no. 3, 513–547.
10. Mark Goresky, Robert Kottwitz, and Robert MacPherson, *Equivariant cohomology, Koszul duality, and the localization theorem*, Invent. Math. **131** (1998), no. 1, 25–83.
11. Hofer Hofer, Kris Wysocki, and Eduard. Zehnder, *Properties of pseudoholomorphic curves in symplectisations. I. Asymptotics*, Ann. Inst. H. Poincaré Anal. Non Linéaire **13** (1996), no. 3, 337–379.
12. Ko Honda, *Transversality theorems for harmonic forms*, Rocky Mountain J. Math. **34** (2004), no. 2, 629–664.
13. Michael Hutchings, *An index inequality for embedded pseudoholomorphic curves in symplectizations*, J. Eur. Math. Soc. (JEMS) **4** (2002), no. 4, 313–361.
14. ———, *Embedded contact homology and its applications*, Proceedings of the International Congress of Mathematicians. Volume II (New Delhi), Hindustan Book Agency, 2010, pp. 1022–1041.
15. Arthur Jaffe and Clifford Taubes, *Vortices and monopoles*, Progress in Physics, vol. 2, Birkhäuser Boston, Mass., 1980, Structure of static gauge theories.
16. András Juhász and Dylan Thurston, *Naturality and mapping class groups in Heegaard Floer homology*, preprint (2012), available at arXiv:1210.4996.
17. Peter B. Kronheimer and Tomasz S. Mrowka, *Monopoles and three-manifolds*, New Mathematical Monographs, vol. 10, Cambridge University Press, Cambridge, 2007.
18. Çağatay Kutluhan, Yi-Jen Lee, and Clifford H. Taubes, *$HF = HM$ I : Heegaard Floer homology and Seiberg–Witten Floer homology*, preprint (2010), available at arXiv:1007.1979.
19. ———, *$HF = HM$ II : Reeb orbits and holomorphic curves for the ech/Heegaard Floer homology correspondence*, preprint (2010), available at arXiv:1008.1595.
20. ———, *$HF = HM$ III : Holomorphic curves and the differential for the ech/Heegaard Floer homology correspondence*, preprint (2010), available at arXiv:1010.3456.

21. ———, $HF = HM$ *IV : The Seiberg–Witten Floer homology/ech correspondence*, preprint (2011), available at arXiv:1107.2297.
22. ———, $HF = HM$ *V : Seiberg–Witten Floer homology and handle addition*, preprint (2012), available at arXiv:1204.0115.
23. Yi-Jen Lee, *Heegaard splittings and Seiberg-Witten monopoles*, Geometry and topology of manifolds, Fields Inst. Commun., vol. 47, Amer. Math. Soc., Providence, RI, 2005, pp. 173–202.
24. Yi-Jen Lee and Clifford Henry Taubes, *Periodic Floer homology and Seiberg-Witten-Floer cohomology*, J. Symplectic Geom. **10** (2012), no. 1, 81–164.
25. Robert Lipshitz, *A cylindrical reformulation of Heegaard Floer homology*, Geom. Topol. **10** (2006), 955–1097 (electronic).
26. Peter Ozsváth and Zoltán Szabó, *Holomorphic disks and topological invariants for closed three-manifolds*, Ann. of Math. (2) **159** (2004), no. 3, 1027–1158.
27. ———, *Holomorphic disks and three-manifold invariants: properties and applications*, Ann. of Math. (2) **159** (2004), no. 3, 1159–1245.
28. ———, *Holomorphic triangles and invariants for smooth four-manifolds*, Adv. Math. **202** (2006), no. 2, 326–400.
29. Nathan Seiberg and Edward Witten, *Electric-magnetic duality, monopole condensation, and confinement in $N = 2$ supersymmetric Yang-Mills theory*, Nuclear Phys. B **426** (1994), no. 1, 19–52.
30. Richard Siefring, *Relative asymptotic behavior of pseudoholomorphic half-cylinders*, Comm. Pure Appl. Math. **61** (2008), no. 12, 1631–1684.
31. Clifford H. Taubes, SW $\Rightarrow$ Gr*: from the Seiberg-Witten equations to pseudo-holomorphic curves*, J. Amer. Math. Soc. **9** (1996), no. 3, 845–918.
32. Clifford H. Taubes, *A compendium of pseudoholomorphic beasts in* $\mathbb{R} \times (S^1 \times S^2)$, Geom. Topol. **6** (2002), 657–814.
33. Clifford H. Taubes, *The Seiberg–Witten equations and the Weinstein conjecture*, Geom. Topol. **11** (2007), 2117–2202.
34. Clifford H. Taubes, *Embedded contact homology and Seiberg-Witten Floer cohomology I*, Geom. Topol. **14** (2010), no. 5, 2497–2581.
35. ———, *Embedded contact homology and Seiberg-Witten Floer cohomology II*, Geom. Topol. **14** (2010), no. 5, 2583–2720.
36. ———, *Embedded contact homology and Seiberg-Witten Floer cohomology III*, Geom. Topol. **14** (2010), no. 5, 2721–2817.
37. ———, *Embedded contact homology and Seiberg-Witten Floer cohomology IV*, Geom. Topol. **14** (2010), no. 5, 2819–2960.
38. ———, *Embedded contact homology and Seiberg-Witten Floer cohomology V*, Geom. Topol. **14** (2010), no. 5, 2961–3000.
39. Clifford Henry Taubes, *Arbitrary N-vortex solutions to the first order Ginzburg-Landau equations*, Comm. Math. Phys. **72** (1980), no. 3, 277–292.
40. Vladimir Turaev, *Torsion invariants of* Spin^c*-structures on* 3*-manifolds*, Math. Res. Lett. **4** (1997), no. 5, 679–695.

Harvard University, Department of Mathematics, One Oxford Street, Cambridge, MA 02138

E-mail address: kutluhan@math.harvard.edu

Proceedings of 19th Gökova
Geometry-Topology Conference
pp. 43 – 88

The combinatorics of Morse theory with boundary

Jonathan M. Bloom

ABSTRACT. We prove several combinatorial results on path algebras over discrete structures related to directed graphs. These results are motivated by Morse theory on a manifold with boundary and, more generally, by Floer theory on a configuration space with boundary. Their purpose is to organize cobordism relationships among moduli spaces in order to define new algebraic invariants. We discuss applications to the Morse and Fukaya categories, and to the work with John Baldwin on a bordered monopole Floer theory.

CONTENTS

Key words and phrases. Morse theory, manifold with boundary, path algebra, hypergraph, monopole Floer homology.

The author was supported by NSF grant DMS-1104032.

1. Introduction

In this article, we prove several combinatorial results on path algebras over discrete structures related to directed graphs. These results are motivated by Morse theory on a manifold with boundary and, more generally, by Floer theory on a configuration space with boundary. Their purpose is to organize cobordism relationships among moduli spaces in order to define new algebraic invariants. Sections 2 through 4 give a fully self-contained exposition of the mathematical content. The Appendix briskly reviews Morse homology on a manifold with boundary, as defined by Kronheimer and Mrowka as the model for monopole Floer homology [9]. We also explain there the connection to cell structures and CW-homology. In the following three subsections of the introduction, we indicate the flavor of the combinatorics, its relation to Morse theory, and applications to Morse and Floer theory with boundary.

1.1. Counting paths

Let $\mathbb{F}$ be the 2-element field. To a transitive digraph G, one can associate a differential graded algebra $\mathcal{A}$ over $\mathbb{F}$, the *path DGA* of G. The $\mathbb{F}$-basis of $\mathcal{A}$ is given by paths in G, the product by concatenation, and the grading by path length. The differential δ of the edge (v_1, v_2) is the sum of all length-two paths from v_1 to v_2, and δ extends to paths by the Leibniz rule. The identity $\delta^2 = 0$ follows from the fact that a sequence of three compatible edges can be concatenated in two orders: $(ab)c$ and $a(bc)$. The element $D \in \mathcal{A}$ given by the sum of all the edges of G satisfies the *structure equation*

$$\delta D = D \circ D$$

as both sides equal the sum of all length-two paths in G. The digraph corresponding to Morse homology on a closed manifold is shown in Figure 1.

In a directed hypergraph, the source and target of an edge are each subsets of the set of vertices. In Section 2, we introduce a class of directed hypergraphs called *higraphs* on which the path DGA is well-defined and the structure equation continues to hold. The higraph underlying the cup product is shown in Figure 2.

In Sections 3 and 4 we extend these results to our main character: the *bigraph* of a higraph (the "b" stands for blow-up and boundary). Each higraph vertex is covered by three bigraph vertices called interior (o), stable (s), and unstable (u), and each higraph edge is covered by a number of bigraph edges called interior and boundary. The grading on the path DGA $\widetilde{\mathcal{A}}$ of a bigraph is defined in terms of weight rather than length. The weight of an interior edge is 1 and the weight of a boundary edge depends on the number and types of vertices in its source and target (Definition 3.7). The differential $\tilde{\delta}$ on a weight w interior edge (resp., boundary edge) is the sum all weight $w + 1$ paths (resp., boundary paths) with the same source and target. Our first main result (Theorem 4.2) verifies that $\tilde{\delta}^2 = 0$. A bigraph is *tree-like* if every edge has singleton target and *rake-like* if every edge has singleton source. For tree-like bigraphs, our second result (Theorem 4.3) states that the element $\tilde{D} \in \widetilde{\mathcal{A}}$ given by the sum of all weight 1 paths between interior

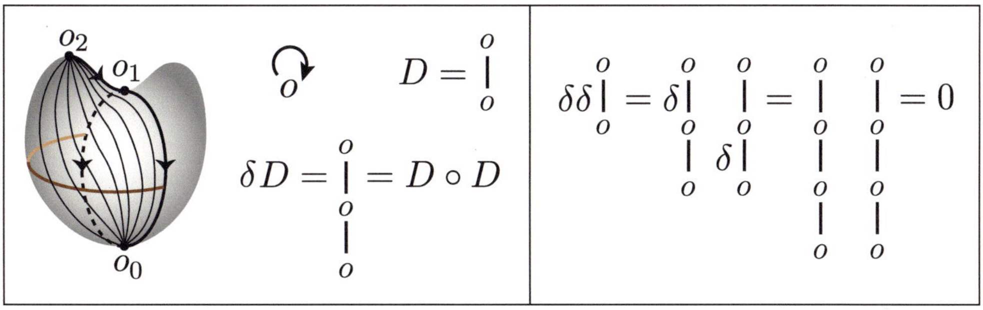

FIGURE 1. The higraph (and digraph) consisting of a single vertex and a single edge corresponds to Morse homology on a closed manifold. The vertex o represents the Morse complex and the edge D represents the Morse differential. The differential δ on the path DGA encodes the key cobordism relation between trajectory spaces illustrated on the surface at left. The structure equation is immediate. On the right-hand side, the identity $\delta^2 = 0$ encodes the fact that each moduli space of twice-broken trajectories bounds two moduli spaces of once-broken trajectories.

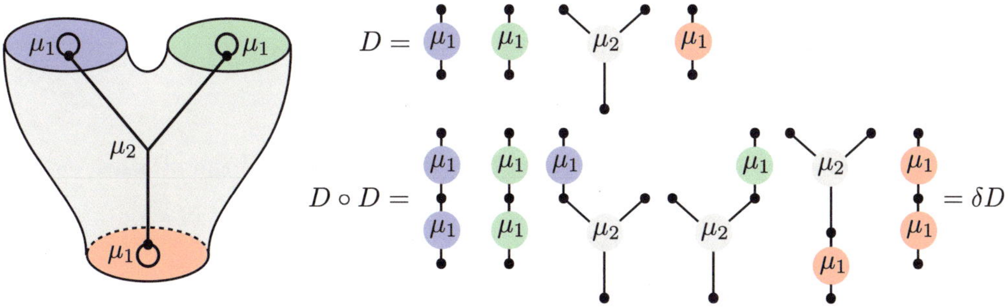

FIGURE 2. The higraph at left consists of three vertices and four edges. The three edges labeled by μ_1 are loops and the fourth edge μ_2 runs from the two top vertices to the bottom vertex. In the context of the Morse category discussed in Section 1.3, the structure equation implies that μ_1 is a differential and μ_2 is a chain map, inducing the cup product structure.

and unstable vertices satisfies the structure equation

$$\tilde{\delta}\tilde{D} = \tilde{D} \circ \tilde{D}.$$

Dually, for rake-like bigraphs, the structure equation holds when $\tilde{D}$ is given by the sum of all weight 1 paths between interior and stable vertices.

The bigraph covering the higraph in Figure 1 is shown at the top of Figure 3. The reader may look to the Appendix, particularly Figure 15, to see how this fundamental example encodes the structure of Morse homology on a manifold with boundary. There are three vertices, four interior edges in black, and four boundary edges in red. The dashed red edge from s to u has weight 0 and the doubled red edge from u to s has weight 2. The other six edges have weight 1. The element D is the sum of all weight 1 paths starting or ending at o or s. The middle portion of Figure 3 shows the value of the differential $\tilde{\delta}$ on each edge, together with a surface illustrating the cobordism relation between trajectory spaces encoded by $\tilde{\delta}(o, o)$. The bottom portion depicts the cancellation of three pairs of paths in $\tilde{\delta}\tilde{D}$ which do not appear in $\tilde{D} \circ \tilde{D}$. The greyscale surfaces along the bottom represent boundary components of 3-manifolds, in order to indicate the relationship of this cancellation to Morse theory. The families of red trajectories lie on the boundary while the families of black trajectories flow into the blue interior. The cancellation reflects the fact that each highlighted broken trajectory arises as the end of a one-parameter family of (broken) trajectories in two ways.

The bigraph in Figure 10 covers the higraph in Figure 2. Now $\tilde{\delta}\tilde{D}$ expands out to 197 paths, 62 of which cancel in pairs, leaving the 135 terms of $\tilde{D} \circ \tilde{D}$. A careful analysis of this cancellation phenomenon underlies the proof of Theorem 4.3.

1.2. The view from Morse theory

The Morse homology package extracts algebraic invariants of smooth manifolds from cobordism relationships between moduli spaces of gradient trajectories. The fundamental cobordism relation takes the schematic form

$$\partial X = X \circ X, \tag{1}$$

in which the boundary of one moduli space is identified with products of other moduli spaces. More precisely, we have

$$\partial \overline{X}_b^a = \bigcup_c X_c^a \times X_b^c. \tag{2}$$

where X_b^a is the moduli space of unparameterized trajectories from a to b, and $\overline{X}_b^a$ denotes a compactification of X_b^a. One defines an endomorphism D on the vector space generated by the set of critical points by setting

$$\langle Da, b\rangle = |X_b^a|.$$

One may also define an endomorphism δD by setting

$$\langle(\delta D)a, b\rangle = |\partial \overline{X}_b^a|.$$

The cobordism relation (2) implies the *structure equation*

$$\delta D = D \circ D, \tag{3}$$

a precise algebraic translation of the schematic (1). Furthermore, δD is the zero map, since $\partial \overline{X}_b^a$ is finite if and only if it is the boundary of a compact 1-manifold and hence

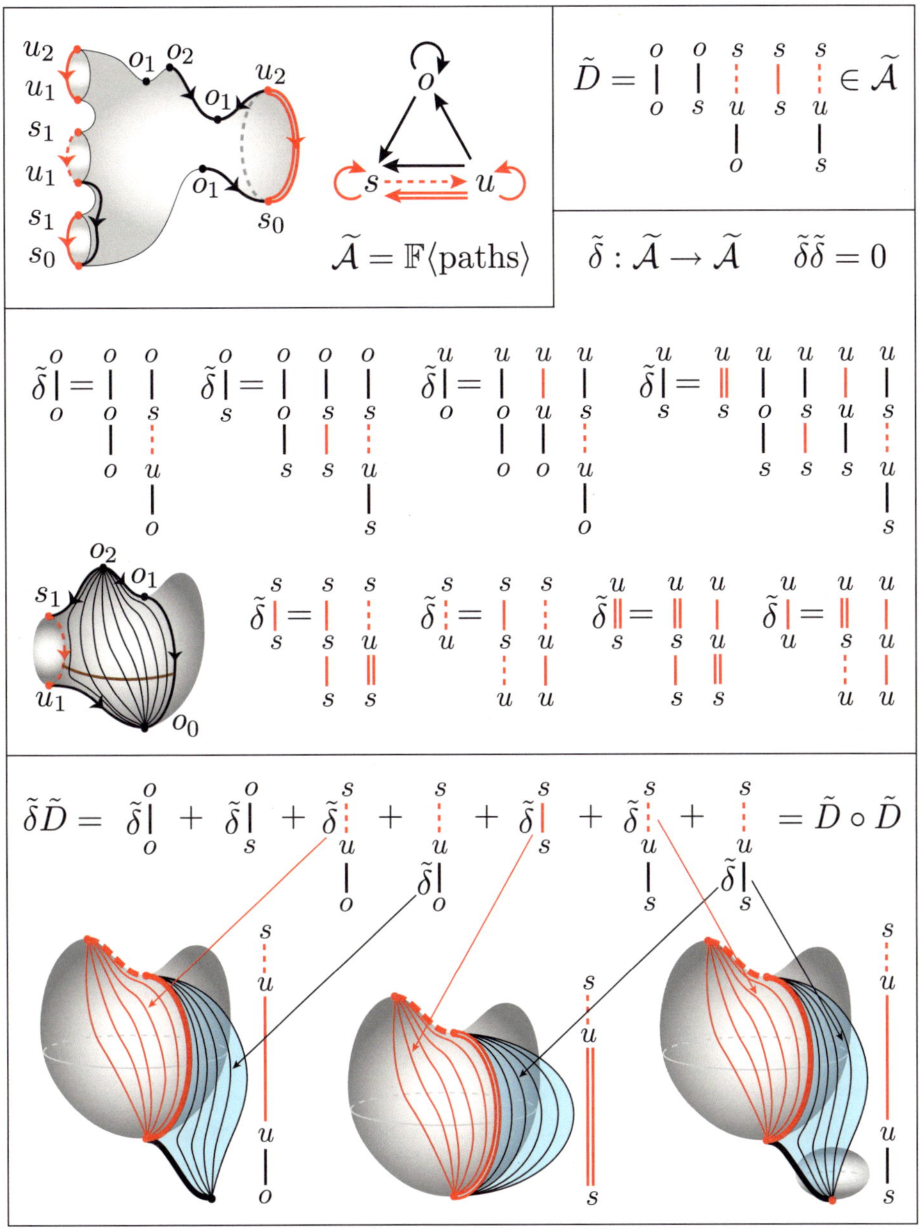

FIGURE 3. The path DGA of the bigraph representing Morse homology on a manifold with boundary (compare with Figure 15).

an even number of points. We conclude from the structure equation that D is indeed a differential. Note that the same framework is used to prove that Morse homology is invariant and functorial. Going one step further, one may define an endomorphism $\delta\delta D$ by setting

$$\langle(\delta\delta D)a, b\rangle = |\partial(\partial\overline{X}_b^a)|.$$

This is the zero map for a different topological reason: the compactification of X_b^a may be given the structure of a smooth manifold with corners.

This viewpoint naturally extends to the case of a manifold with boundary.

For a manifold with boundary, the space X is itself a union of products of moduli spaces of trajectories (that is, X is composed of several types of possibly-broken trajectories). The space $\overline{X}$ is built from the compactifications of the factors, its boundary given by the Leibniz rule. The structure equation continues to hold, although now after the cancellation of several pairs of terms in δD which do not appear in $D \circ D$ as illustrated in Figure 3. And the maps δD and $\delta\delta D$ still vanish. We can summarize the situation as follows: without boundary, Morse homology depends on geometric analysis and a bit of topology; with boundary, Morse homology depends on (more involved) geometric analysis, a bit of topology, and the simplest case of the combinatorics that is comprehensively developed herein.

Remark 1.1. Kronheimer and Mrowka study Morse theory with boundary in the context of monopole Floer homology. They do not prove that $\overline{X}$ is a manifold with corners, but rather that it is stratified by smooth manifolds and has an even number of ends in the 1-dimensional case. The example in Figure 11 indicates that $\overline{X}$ is not a smooth manifold with corners when its dimension is three or more. On the other hand, Proposition 4.8 shows that $|\partial(\partial\overline{X})| = 0$, which corresponds to the identity $\delta^2 = 0$ in the path DGA setting and suggests that $\overline{X}$ may be a manifold with corners in the 2-dimensional case.

1.3. Morse and Floer theory with boundary

Morse homology is an approach to analyzing the topology of a smooth manifold by studying the gradient flow of a generic smooth functional. Floer theory generalizes this approach to infinite-dimensional configuration spaces arising in symplectic geometry and gauge theory. Two important examples, Lagrangian intersection Floer homology and monopole Floer homology, may be placed in parallel with finite-dimensional Morse homology as follows:

- The *Morse complex* $C(f)$ of a smooth function f on a closed manifold is generated by the critical points of f; the differential counts gradient trajectories between critical points.
- The *Langrangian Floer complex* $CF(L_0, L_1)$ of two Lagrangian submanifolds of a symplectic manifold X is generated by the points of $L_0 \cap L_1$; the differential counts pseudo-holomorphic bigons with edge i on L_i. Here the functional is the symplectic action functional on the space of paths from L_0 to L_1 [8].

- The *monopole Floer complex* $\widehat{C}(Y)$ of a closed 3-manifold Y is generated by solutions (called monopoles) to the 3-dimensional Seiberg-Witten equations on Y; the differential counts solutions to the 4-dimensional Seiberg-Witten equations on $Y \times \mathbb{R}$. Here the functional is the Chern-Simons-Dirac functional on the configuration space $\mathcal{B}(Y)$ consisting of pairs of certain connections and spinors over Y modulo gauge [9].

We caution that these rough descriptions suppresses auxiliary data (such as metrics, perturbations, and almost-complex structures) and ignore bubbling phenomena that may prevent the Langrangian Floer complex from being well-defined. Note also that the configuration space $\mathcal{B}(Y)$ has a boundary consisting of reducible configurations, which is why there are three versions of monopole Floer homology. The correspondence with Morse homology is reflected by the exact sequences

$$\cdots \longrightarrow H_*(\partial M) \longrightarrow H_*(M) \longrightarrow H_*(M, \partial M) \longrightarrow \cdots$$
$$\cdots \longrightarrow \overline{HM}_\bullet(Y) \longrightarrow \widecheck{HM}_\bullet(Y) \longrightarrow \widehat{HM}_\bullet(Y) \longrightarrow \cdots .$$

Singular cohomology admits an associative product structure, the cup product, which may be defined Morse-theoretically using gradient trees (or equivalently, triple intersections of stable and unstable manifolds with respect to several Morse functions). Fukaya recognized that deeper multiplicative structure is naturally encoded by an A_∞-category, which determines the Massey products and rational homotopy type [7].

- The objects of the *Morse category* of a closed manifold are smooth functions and $\mathrm{Mor}(f_0, f_1) = C(f_0 - f_1)$. The multiplication map

$$\mu_k : C(f_0 - f_1) \otimes \cdots \otimes C(f_{k-1} - f_k) \to C(f_0 - f_k)$$

  counts gradient trees with k inputs and 1 output.

Here μ_1 is the Morse differential and μ_2 is the chain map defined by

$$\langle \mu_2(a \otimes b), c \rangle = |\{x \in M : \lim_{t \to \infty} \phi_t^{f_0 - f_1}(x) = a, \ \lim_{t \to \infty} \phi_t^{f_1 - f_2}(x) = b, \ \lim_{t \to \infty} \phi_t^{f_0 - f_2}(x) = c\}|$$

where $\phi_t^f : M \to M$ denotes the flow induced by ∇f. The map μ_3 provides a homotopy between $\mu_2(\mu_2(\cdot, \cdot), \cdot)$ and $\mu_2(\cdot, \mu_2(\cdot, \cdot))$. In particular, μ_2 induces an associative product on homology, dual to the cup product. See [15] for background on A_∞-categories.

One application of our combinatorial results is to extend the definition of the Morse category to manifolds with boundary. For the appropriate sequence of higraphs, the path DGAs of bigraphs specify the higher multiplications, and the structure equation is equivalent to the A_∞-relations. For instance, the map μ_2 is defined to count broken gradient trees as specified in Figure 10. The higraph for μ_3 is shown in Figure 5, and the general case is similar.

Fukaya introduced the Morse category as a model for his eponymous category, which encodes the higher multiplicative structure of Langrangian intersection Floer homology.

- The objects of the *Fukaya category* of a symplectic manifold X are Lagrangian submanifolds and $\mathrm{Mor}(L_0, L_1) = CF(L_0, L_1)$. The multiplication map
$$\mu_k^{\mathrm{Lag}} : CF(L_0, L_1) \otimes \cdots \otimes CF(L_{k-1}, L_k) \to CF(L_0, L_k)$$
counts pseudo-holomorphic $(k+1)$-gons with edge i on L_i.

The connection with the Morse category goes as follows. Given a smooth function f on M, the graph of df is a Lagrangian in T^*M with the standard symplectic structure. The Fukaya category generated by such Lagrangians is A_∞-equivalent to the Morse category of M, which implies the Arnold conjecture for the zero-section [7], [1].

The Fukaya category is a central algebraic structure in symplectic geometry and mirror symmetry. It is also notoriously difficult to define or compute rigorously [8], [15]. See [13] for a state-of-the-art computation: the Fukaya category of the 2-dimensional torus. A divide-and-conquer approach to computation might go as follows: chop the symplectic manifold into simpler pieces and reassemble its Fukaya category by algebraically compiling the data of some notion of Fukaya category for each piece. Akaho has taken a step in this direction, defining the Floer complex $CF(L_1, L_2)$ of two Lagrangians in an open symplectic manifold with concave end [2]. A slice of the end inherits a contact structure and intersects the Lagrangians in Legendrians. The complex $CF(L_1, L_2)$ is generated by both points of $L_1 \cap L_2$ and Reeb chords between the Legendrians. The Reeb chords play the role of boundary critical points and the differential takes the same form as the Morse differential on a manifold with boundary. As such, our bigraph framework should specify the higher multiplications needed to define the Fukaya category of an open symplectic manifold with concave end as well.

Our primary application, in joint work with John Baldwin, is to define a gauge-theoretic analogue of the Fukaya category using monopole Floer theory. This monopole category combines older ideas of Segal, Donaldson, and Fukaya [6] with the comprehensive analysis of Kronheimer and Mrowka [9].

- The objects of the *monopole category* of a closed surface Σ are 3-manifolds with boundary parameterized by Σ, and $\mathrm{Mor}(Y_0, Y_1) = \widehat{C}(Y_0 \cup_\Sigma -Y_1)$. The map
$$\mu_k^{\mathrm{mon}} : \widehat{C}(Y_0 \cup -Y_1) \otimes \cdots \otimes \widehat{C}(Y_{k-1} \cup -Y_k) \to \widehat{C}(Y_0 \cup -Y_k)$$
counts monopoles on a 4-dimensional cobordism
$$W : (Y_0 \cup -Y_1) \sqcup \cdots \sqcup (Y_{k-1} \cup -Y_k) \to Y_0 \cup -Y_k$$
over a family of metrics parameterized by the $(k-2)$-dimensional associahedron.

The cobordism W is constructed by gluing $\partial Y_i \times [0,1] \subset Y_i \times [0,1]$ to $\Sigma \times e_i \subset \Sigma \times \triangle$ for $0 \le i \le k$, where $\triangle$ is a polygonal region bounded by $k+1$ line segments alternating with $k+1$ arcs e_i. The case $k = 2$ is depicted in Figure 4. In general, the family of metrics interpolates between all ways of cutting W along disjoint hypersurfaces $Y_i \cup -Y_j$ corresponding to chords in $\triangle$ between non-adjacent arcs e_i. Figure 5 illustrates the interval of metrics in the case $k = 3$ from a less rigid perspective.

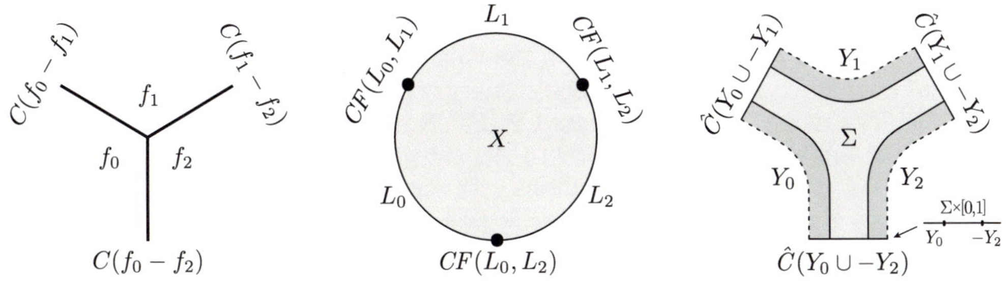

FIGURE 4. The μ_2 multiplication map in the Morse, Fukaya, and monopole A_∞-categories counts gradient trees, pseudo-holomorphic triangles, and monopoles on a 4-dimensional cobordism, respectively. On homology, this map induces composition of morphisms in an ordinary category.

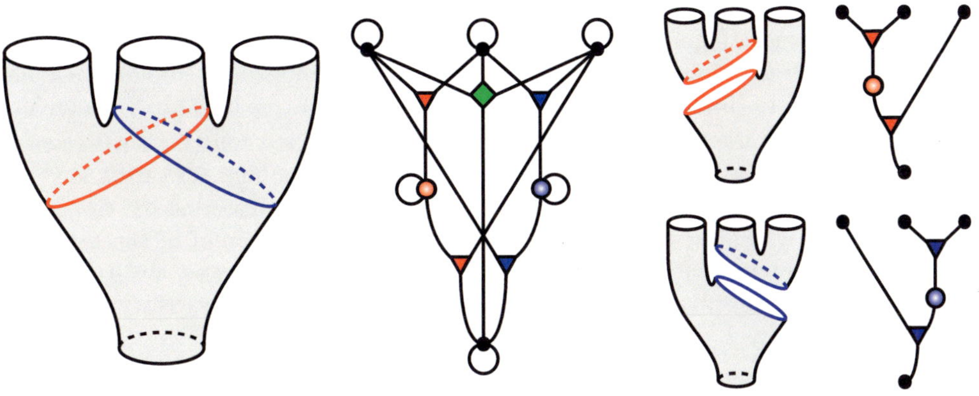

FIGURE 5. The μ_3 multiplication map is represented by the green edge from the three top vertices to the bottom vertex. The full higraph contains 6 vertices and 11 edges. The compositions $\mu_2(\mu_2(\cdot,\cdot),\cdot)$ and $\mu_2(\cdot,\mu_2(\cdot,\cdot))$ are each represented by paths of length 2 at right. Due to boundary (reducibles), the terms of the monopole maps μ_1^{mon}, μ_2^{mon}, μ_3^{mon} are encoded by the corresponding bigraph. Roughly, μ_3^{mon} counts monopoles on a 4-dimensional cobordism (at left) over a family of metrics parameterized by the 1-dimensional associahedron: the interval $[-1,1]$. The metric stretches normal to the red (resp., blue) hypersurface as the parameter goes to -1 (resp., 1). The degenerate metrics at ± 1 correspond to composite cobordisms (at right). The structure equation implies the A_∞-relation for μ_3^{mon} (not shown).

In light of the isomorphism between monopole and Heegaard Floer homology (see [4] and [11]), it is natural to conjecture that the monopole category of a surface Σ of genus g is A_∞-equivalent to the subcategory of the Fukaya category of $\mathrm{Sym}^g(\Sigma)$ generated by Heegaard tori. Furthermore, the monopole category fits into the framework of a *bordered monopole Floer theory*, a form of extended TQFT. Roughly, the bordered monopole Floer package is a 2-functor sending surfaces to A_∞-categories, 3-dimensional cobordisms to A_∞-functors, and 4-dimensional cobordisms with corners to A_∞-natural transformations. In dimensions 2 and 3, this theory is analogous to the version of bordered Heegaard Floer theory developed by Lekili and Perutz [12] (which in turn is related to the original version [14]). In our case, for each genus, the monopole category is split-generated by an explicit, finite set of parameterized handlebodies through the surgery exact triangle, and the homology algebra is known. We hope that this theory will point toward an axiomatization of Floer homology in low-dimensional topology.

The preceding applications utilize the sequence of higraphs arising from associahedra encoding homotopy associativity. Other higraphs are useful as well. The link surgery spectral sequence is most naturally defined using permutohedra encoding homotopy commutativity; these higraphs have the vertex set of a hypercube with an edge between every ordered pair of vertices [3]. Baldwin and the author's constructions in bordered monopole Floer theory make use of more general polytopes called graph associahedra [5], including cyclohedra related to Hochschild homology. More basically, a tree higraph (one non-loop edge with singleton target) leads to an $\widehat{HM}_\bullet$ map for a connected cobordism with disconnected source as in Figure 10. Dually, a rake higraph (one non-loop edge with singleton source) leads to an $\widetilde{HM}_\bullet$ map for a connected cobordism with disconnected target. As a final example, we have appended the higraph organizing the proof of the monopole surgery exact triangle as Figure 19. To cover these and future cases, we have aimed for a definition of higraphs that is as general as possible while still satisfying a unified combinatorial theory.

2. Higraphs

Definition 2.1. A *directed graph*, or *digraph*, $G = (V, E)$ consists of a vertex set V and an edge set $E \subset V \times V$. The edge $e = (v_1, v_2)$ has *source* $s(e) = v_1$ and *target* $t(e) = v_2$.

A *directed hypergraph* $\mathcal{H} = (\mathcal{V}, \mathcal{E})$ consists of a vertex set $\mathcal{V}$ and an edge set $\mathcal{E} \subset 2^{\mathcal{V}} \times 2^{\mathcal{V}}$. The edge $\varepsilon = (I, J)$ has *source* $s(\varepsilon) = I \subset \mathcal{V}$ and *target* $t(\varepsilon) = J \subset \mathcal{V}$.

We reserve the term higraph for a special kind of directed hypergraph.

Definition 2.2. A *higraph* $\mathcal{H}$ is a directed hypergraph $(\mathcal{V}, \mathcal{E})$ such that the following properties hold for all edges $\varepsilon, \varepsilon' \in \mathcal{E}$.

TRANSITIVE: If $t(\varepsilon) \cap s(\varepsilon')$ is non-empty, then $t(\varepsilon) \cap s(\varepsilon') = \{K\}$ for some $K \in \mathcal{V}$ and

$$(s(\varepsilon) \cup (s(\varepsilon') - \{K\}), (t(\varepsilon) - \{K\}) \cup t(\varepsilon')) \in \mathcal{E} \tag{4}$$

where each of the above unions is of disjoint sets.

ACYCLIC: If $t(\varepsilon) \cap s(\varepsilon')$ and $t(\varepsilon') \cap s(\varepsilon)$ are both non-empty, then $\varepsilon = \varepsilon' = (\{K\}, \{K\})$ for some $K \in \mathcal{V}$.

An edge of the form $(\{K\}, \{K\})$ is called a *loop*. Transitivity ensures that adjacent edges ε and ε' may be glued to form a single edge (4) in a well-defined manner. Acyclicity ensures that the only edges with overlapping source and target are loops, and that loops cannot result from gluing distinct adjacent edges.

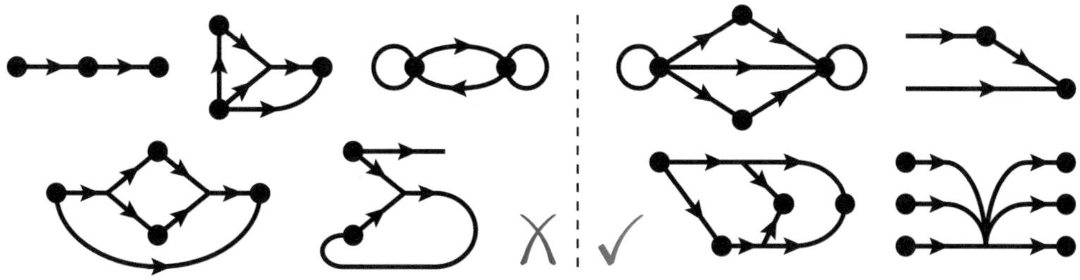

FIGURE 6. Of the nine directed hypergraphs above, only the four at right are higraphs. All vertices are shown as black dots. The bottom-right higraph has only one edge; the source and target of this edge each consist of three vertices. Two edges in the top-right higraph have empty source; gluing the upper two edges results in the lower edge.

2.1. Paths

We recall the notion of a directed tree.

Definition 2.3. A *directed tree* T is a digraph which is acyclic as an undirected graph. For clarity, we will refer to the vertices and edges of T as *nodes* and *arrows*, respectively.

A path in a higraph has the structure of a directed tree – the nodes index the edges of the path, while the arrows index the breaks between edges.

Definition 2.4. A *path* $\rho = (T, \rho)$ in a higraph $\mathcal{H} = (\mathcal{V}, \mathcal{E})$ consists of a non-empty, directed tree $T = (V, E)$ together with a map $\rho : V \to \mathcal{E}$ such that:

(1) The map ρ induces a well-defined map $\rho : E \to \mathcal{V}$ such that

$$\{\rho(e)\} = t(\rho(s(e))) \cap s(\rho(t(e))).$$

(2) If distinct arrows e and e' have the same source or the same target, then we have $\rho(e) \neq \rho(e')$.

Paths ρ and ρ' are *isomorphic* if there is a digraph isomorphism $f : T \to T'$ of the associated directed trees such that $\rho = \rho' f$.

In fact, the automorphism group of a path is trivial. This is clear if ρ covers a loop. If ρ does not cover a loop, then transitivity and acyclicity imply that the leaves of T are mapped to distinct edges. So in this case, any automorphism of ρ must fix the leaves of T, and is therefore the identity. Thus, paths are unique up to unique isomorphism, and we consider isomorphic paths to be the same.

Definition 2.5. Let $\mathcal{P}$ denote the set of paths in $\mathcal{H}$. Given a path ρ, the *length* of ρ is the number of edges it contains, given by $l(\rho) = |V|$. We refer to the arrows of T as *breaks*. The number of breaks in ρ is given by $b(\rho) = |E| = l(\rho) - 1$. The *source* of a path ρ is the subset of $\mathcal{V}$ given by

$$s(\rho) = \bigcup_{v \in V} \left(s(\rho(v)) - \{\rho(e) \mid t(e) = v\}\right).$$

The *target* of a path ρ is the subset of $\mathcal{V}$ given by

$$t(\rho) = \bigcup_{v \in V} \left(t(\rho(v)) - \{\rho(e) \mid s(e) = v\}\right).$$

Transitivity implies that the union defining $s(\rho)$ is of disjoint sets, and similarly for $t(\rho)$. Thus each vertex $K \in s(\rho)$ is *sourced* by a unique node $v \in V$. Similarly, each vertex $K \in t(\rho)$ is *targeted* by a unique node $v \in V$.

2.2. Path operations

We next define several ways to combine or modify paths. *Concatenation* joins two compatible paths together. *Swapping* cuts out a subpath of a path and replaces it with another compatible path. *Refinement* and *gluing* are special cases of swapping. Refinement expands a single edge to a compatible path, while gluing contracts a subpath to a single edge.

2.2.1. Concatenation

Definition 2.6. The partially-defined *concatenation* map $c : \mathcal{P} \times \mathcal{P} \dashrightarrow \mathcal{P}$ joins compatible paths together. Consider paths ρ_1 and ρ_2 with associated trees $T_1 = (V_1, E_1)$ and $T_2 = (V_2, E_2)$. The path $\rho = c(\rho_1, \rho_2)$ is defined precisely when $t(\rho_1) \cap s(\rho_2)$ is non-empty, in which case the intersection contains a single vertex $K \in \mathcal{V}$ targeted by $v_1 \in V_1$ and sourced by $v_2 \in V_2$. The directed tree $T = (V, E)$ associated to ρ is given by

$$V = V_1 \cup V_2$$

and

$$E = E_1 \cup E_2 \cup \{(v_1, v_2)\}.$$

The map $\rho : V \to \mathcal{E}$ is given by $\rho_1 \sqcup \rho_2$. The induced map $\rho : E \to \mathcal{V}$ sends (v_1, v_2) to K.

2.2.2. Swapping

Let $\mathcal{Q} = \{(\rho, T') \mid \rho = (T, \rho) \in \mathcal{P},\ T' \text{ is a non-empty subtree of } T\}$.

Definition 2.7. The partially-defined *swapping map* $w : \mathcal{Q} \times \mathcal{P} \dashrightarrow \mathcal{P}$ replaces a subpath of a path with a compatible path. Given a path $\rho_1 = (T_1, \rho_1)$ and non-empty subtree $T' = (V', E')$ of T_1, we obtain a subpath $\rho' = (T', \rho')$ by setting $\rho' = \rho_1|_{V'}$. The path $\rho_2 = w((\rho_1, T'), \rho'')$ is defined precisely when $s(\rho') = s(\rho'')$ and $t(\rho') = t(\rho'')$. In this case, the associated tree $T_2 = (V_2, E_2)$ has node set given by

$$V_2 = (V_1 - V') \cup V''.$$

To describe the arrow set E_2, note that for each arrow $e \in (V_1 - V') \times V'$, the vertex $\rho'(e) \in s(\rho'')$ is sourced by a unique node $t''(e)$ of T''. Similarly, for each arrow $e \in V' \times (V_1 - V')$, the vertex $\rho'(e) \in t(\rho'')$ is targeted by a unique node $s''(e)$ of T''. We set

$$E_2 = (E_1 \cap ((V_1 - V') \times (V_1 - V'))) \cup E_s \cup E_t \cup E'' \subset V_2 \times V_2$$

where

$$E_s = \{(s(e), t''(e)) \mid e \in (V_1 - V') \times V'\}$$

and

$$E_t = \{(s''(e), t(e)) \mid e \in V' \times (V_1 - V')\}.$$

The map $\rho_2 : V_2 \to \mathcal{E}$ is induced by ρ_1 on $V_1 - V'$ and by ρ'' on V''. Note that ρ_2 has the same source and target as the original path ρ_1.

Since swapping preserves source and target, swapping and concatenation commute in the following sense. Fix paths ρ_1, ρ_2, and ρ'', and let T_1' and T_2' be non-empty subtrees of T_1 and T_2, respectively. Then

$$w((c(\rho_1, \rho_2), T_1'), \rho'') = c(w((\rho_1, T_1'), \rho''), \rho_2) \qquad (5)$$

$$w((c(\rho_1, \rho_2), T_2'), \rho'') = c(\rho_1, w((\rho_2, T_2'), \rho'')) \qquad (6)$$

where the left-hand side of each equation is defined if and only if the right-hand side is defined.

Of course, if we swap out a subpath and then swap it back in, we end up where we started. More precisely, if we define the map $W : \mathcal{Q} \times \mathcal{P} \to \mathcal{Q} \times \mathcal{P}$ by

$$W((\rho_1, T'), \rho'') = ((w((\rho_1, T'), \rho''), T''), \rho'), \qquad (7)$$

then $W \circ W$ is the identity map. We are most interested in the following two special cases of swapping, which are mutual inverses in the sense of (7).

2.2.3. Refinement

We first suppose that the subpath ρ' consists of a single edge $\varepsilon^* = \rho'(v^*)$. Let

$$\mathcal{R} = \{(\rho, v^*) \mid \rho = (T, \rho) \in \mathcal{P},\ v^* \text{ is a vertex of } T\}.$$

Definition 2.8. The partially-defined *refinement map* $r : \mathcal{R} \times \mathcal{P} \dashrightarrow \mathcal{P}$ replaces a single edge $\varepsilon^* = \rho_1(v^*)$ of a path ρ_1 with a compatible path ρ'':

$$r((\rho_1, v^*), \rho'') = w((\rho_1, v^*), \rho'').$$

2.2.4. Gluing

We next suppose that the substituted path ρ'' consists of a single edge ε^*.

Definition 2.9. The *gluing map* $q : \mathcal{Q} \to \mathcal{P}$ replaces a subpath ρ' of a path ρ_1 with the single edge $\varepsilon^* = (s(\rho'), t(\rho'))$:

$$q(\rho_1, T') = w((\rho_1, T'), \varepsilon^*).$$

Note that ε^* is indeed an edge of $\mathcal{H}$ by transitivity.

By definition, the tree T_2 underlying $\rho_2 = q(\rho_1, T')$ results from contracting the subtree T' of T_1 to a single vertex v^*, with $\rho_2(v^*) = \varepsilon^*$. In particular, the projection map $\pi : V_1 \to V_2$ given by

$$\pi(v) = \begin{cases} v & \text{if } v \in V_1 - V', \\ v^* & \text{if } v \in V'. \end{cases} \tag{8}$$

induces a bijection $\pi : (E_1 - E') \to E_2$ between the arrows of T_1 that are outside T' and the arrows of T_2.

Definition 2.10. Let e be a break in a path ρ. We use the notation $q(\rho, e)$ to denote the result of gluing the path ρ along the one-arrow subtree induced by e. Note that gluing a path at a break reduces length by 1.

Definition 2.11. A *k-step gluing sequence* on a path ρ_1 is a k-tuple $(e_1, \ldots, e_k)$ with e_i a break on ρ_i with $\rho_{i+1} = q(\rho_i, e_i)$ for $1 \leq i \leq k$. We may write this as

$$\rho_1 \xrightarrow{e_1} \rho_2 \xrightarrow{e_2} \cdots \xrightarrow{e_k} \rho_{k+1}$$

to emphasize the paths along the way. An isomorphism of k-step gluing sequences is a sequence of isomorphisms between the corresponding paths which identify the corresponding breaks.

Gluing sequences are unique up to unique isomorphism, so we consider isomorphic gluing sequences to be the same. The following lemma is established in Section 4.6.1.

Lemma 2.1. *The number of k-step gluing sequences between any two fixed paths in a higraph is a multiple of $k!$.*

2.3. Path DGA

The path algebra on a digraph is associative but not commutative. The following generalization to higraphs is neither associative nor commutative in general.

Definition 2.12. The *path DGA* of the higraph $\mathcal{H}$ over the field $\mathbb{F}$ consists of a graded algebra $\mathcal{A}$ equipped with a differential $\delta : \mathcal{A} \to \mathcal{A}$. The algebra $\mathcal{A}$ has underlying vector space $\mathbb{F}\langle\mathcal{P}\rangle$, with the subspace $\mathcal{A}^k$ in grading k spanned by the set $\mathcal{P}^k$ of paths of length k. Multiplication is bilinear and given on paths by $\rho_1\rho_2 = c(\rho_1, \rho_2)$ if defined, and $\rho_1\rho_2 = 0$ otherwise.

The differential δ is determined by the Leibniz rule and linearity once we specify its value on each edge $\varepsilon \in \mathcal{E}$ regarded as an element of $\mathcal{A}$. The value of δ on an edge ε equals the sum of all refinements of ε to a path of length 2. In particular, the differential has degree 1.

Since a path ρ may often be expressed as a fully-associated product of edges in several ways, we should take care that δ is well-defined. This issue is resolved by noting that we may define δ directly on a path $\rho = ((V, E), \rho)$ by

$$\delta\rho = \sum_{v\in V} \sum_{\rho'\in\mathcal{P}^2} r((\rho, v), \rho') \tag{9}$$

where $r((\rho, v), \rho')$ is defined to be zero on invalid input. This definition is consistent with our earlier definition of $\delta\varepsilon$, and the Leibniz rule now follows from (5) and (6):

$$\begin{aligned}
\delta(\rho_1\rho_2) &= \sum_{v\in V} \sum_{\rho'\in\mathcal{P}^2} r((\rho_1\rho_2, v), \rho') \\
&= \sum_{v\in V_1} \sum_{\rho'\in\mathcal{P}^2} r((\rho_1\rho_2, v), \rho') + \sum_{v\in V_2} \sum_{\rho'\in\mathcal{P}^2} r((\rho_1\rho_2, v), \rho') \\
&= \left(\sum_{v\in V_1} \sum_{\rho'\in\mathcal{P}^2} r((\rho_1, v), \rho')\right) \rho_2 + \rho_1 \left(\sum_{v\in V_2} \sum_{\rho'\in\mathcal{P}^2} r(\rho_2, v), \rho')\right) \\
&= (\delta\rho_1)\rho_2 + \rho_1(\delta\rho_2).
\end{aligned}$$

To see that δ is a differential, it is useful to introduce its adjoint ∂ with respect to the symmetric, bilinear pairing

$$\langle\,\cdot\,,\,\cdot\,\rangle : \mathcal{A} \times \mathcal{A} \to \mathbb{F}$$

with $\langle\rho_1, \rho_2\rangle = 1$ if and only if $\rho_1 = \rho_2$.

Definition 2.13. The map $\partial : \mathcal{A} \to \mathcal{A}$ is defined on the path ρ with underlying tree $T = (V, E)$ by

$$\partial\rho = \sum_{e\in E} q(\rho, e).$$

Proposition 2.2. *We have*

$$\langle\partial a, b\rangle = \langle a, \delta b\rangle$$

for all $a, b \in \mathcal{A}$. Furthermore, $\partial^2 = 0$ and $\delta^2 = 0$.

Proof. For the first statement, by linearity it suffices to verify that

$$\langle \partial\rho_1, \rho_2\rangle = \langle \rho_1, \delta\rho_2\rangle.$$

The left-hand side counts the number of 1-step gluing sequences $\rho_1 \xrightarrow{e_1} \rho_2$. The right-hand side counts all ways to refine an edge of ρ_2 to a path of length 2 to obtain ρ_1. Equality follows from the fact that gluing and refinement are inverses as in (7).

To prove $\partial^2 = 0$, it suffices to show that $\langle \partial^2\rho_1, \rho_3\rangle$ vanishes for all paths ρ_1 and ρ_3. Indeed, the term $\langle \partial^2\rho_1, \rho_3\rangle$ counts the number of 2-step gluing sequences $\rho_1 \xrightarrow{e_1} \rho_2 \xrightarrow{e_2} \rho_3$, which is even by Lemma 2.1.

Now $\delta^2 = 0$ because $\langle \delta^2 a, b\rangle = \langle \delta a, \partial b\rangle = \langle a, \partial^2 b\rangle = \langle a, 0\rangle = 0$. □

Definition 2.14. We define the element $D \in \mathcal{A}$ by

$$D = \sum_{\varepsilon\in\mathcal{E}} \varepsilon = \sum_{\rho\in\mathcal{P}^1} \rho.$$

Proposition 2.3. *The element D satisfies the structure equation*

$$\delta D = D \circ D.$$

Proof. We claim

$$\delta D = \sum_{\rho\in\mathcal{P}^2} \rho = D \circ D.$$

The left-hand equality holds because each path ρ of length 2 arises as the refinement of a unique edge $\varepsilon = (s(\rho), t(\rho))$. The right-hand equality holds because each path of length 2 arises as the product of two (not necessarily distinct) edges in a unique way. □

3. Abstract paths

In Section 4, we will define the notion of the bigraph $\widetilde{\mathcal{H}}$ associated to a higraph $\mathcal{H}$. We will then systematically extend the arguments and results of Section 2 to bigraphs. In this section, we assemble the necessary lemmas, which concern the abstract form of a path in a bigraph. We call such a form an abstract path, as it is defined without reference to a particular bigraph. We now make this notion precise.

Definition 3.1. An *abstract vertex* is an element of the set $\mathcal{V}_{ab} = \{o, u, s\}$. The symbols o, u, and s stand for *interior*, *unstable*, and *stable*, respectively. Both unstable and stable abstract vertices are considered *boundary*.

An *abstract edge* is an element

$$\varepsilon = (\mu, (o_1, u_1, s_1), (o_2, u_2, s_2))$$

of the subset

$$\mathcal{E}_{ab} \subset \mathbb{F} \times (\mathbb{N} \times \mathbb{N} \times \mathbb{N}) \times (\mathbb{N} \times \mathbb{N} \times \mathbb{N}).$$

This subset is cut out by the following two conditions:

Condition 0: If $\mu = 0$ then $s_1 = u_2 = 0$.
Condition 1: If $\mu = 1$ then $o_1 = o_2 = 0$.

An abstract edge ε is *interior* if $\mu = 0$ and *boundary* if $\mu = 1$.

We think of the abstract edge ε as having $o_1+u_1+s_1$ incoming ends (or source vertices), of which o_1 are interior, u_1 are unstable, and s_1 are stable. Similarly, ε has $o_2 + u_2 + s_2$ outgoing ends, of which o_2 are interior, u_2 are unstable, and s_2 are stable. Condition 0 says that the incoming ends of an interior abstract edge are interior or unstable, while the outgoing ends are interior or stable. Condition 1 says that the ends of a boundary abstract edge are boundary abstract vertices.

We can assemble abstract edges into an abstract path, so long as there are sufficiently many ends for concatenation.

Definition 3.2. An *abstract path* is a triple $\tau = (T, \tau, \sigma)$ consisting of a non-empty directed tree T and maps

$$\tau : V \to \mathcal{E}_{ab}$$
$$\sigma : E \to \mathcal{V}_{ab}$$

such that the following hold for each $v \in V$, with $\tau(v) = (\mu, (o_1, u_1, s_1), (o_2, u_2, s_2))$:

$$\begin{aligned} o_1 &\geq |\{e \in E \,|\, t(e) = v,\ \sigma(e) = o\}|, \\ u_1 &\geq |\{e \in E \,|\, t(e) = v,\ \sigma(e) = u\}|, \\ s_1 &\geq |\{e \in E \,|\, t(e) = v,\ \sigma(e) = s\}|, \\ o_2 &\geq |\{e \in E \,|\, s(e) = v,\ \sigma(e) = o\}|, \\ u_2 &\geq |\{e \in E \,|\, s(e) = v,\ \sigma(e) = u\}|, \\ s_2 &\geq |\{e \in E \,|\, s(e) = v,\ \sigma(e) = s\}|. \end{aligned}$$

An abstract path τ is *boundary* if all of its abstract edges are boundary.

We interpret an arrow e from node v_1 to node v_2 to signify that the abstract edges $\tau(v_1)$ and $\tau(v_2)$ are concatenated at an abstract vertex of type $\sigma(e)$. In this sense, an abstract path has a well-defined number of source and target abstract vertices of each type. For example, the number of incoming stable ends is given by

$$s_1(\tau) = \left(\sum_{v \in V} s_1(v) \right) - |\sigma^{-1}(s)|,$$

while the number of outgoing unstable ends is given by

$$u_2(\tau) = \left(\sum_{v \in V} u_2(v) \right) - |\sigma^{-1}(u)|.$$

Definition 3.3. An abstract path τ is called *legal* if $s_1(\tau) = u_2(\tau) = 0$; equivalently, τ is legal if its ends are compatible with an interior abstract edge (see Condition 0 of Definition 3.1).

It is convenient to pull back the terminology in Definition 3.1, so that an arrow $e \in E$ is interior, unstable, or stable according to the value of $\sigma(e)$. Similarly, a node $v \in V$ is interior or boundary according to the value of $\mu(v)$. We may decompose the node set V as $V^0 \cup V^1$ where $V^0 = \mu^{-1}(0)$ denotes the set of interior nodes and $V^1 = \mu^{-1}(1)$ denotes the set of boundary nodes.

Definition 3.4. The *interior neighborhood* of a node $v \in V$ is the set of adjacent interior nodes:

$$\mathrm{Nbhd}^o(v) = V^o \cap \{v' \in V \,|\, (v, v') \in E \text{ or } (v', v) \in E\}.$$

The *interior star* of v is the subtree $\mathrm{Star}^o(v)$ of T induced by $\{v\} \cup \mathrm{Nbhd}^o(v)$.

3.1. Gluing moves on abstract paths

We have introduced all the necessary terminology to define the gluing moves on abstract paths. We illustrate these moves in Figure 7, while precise definitions are given below.

Definition 3.5. The three gluing moves on an abstract path $\tau = (T, \tau, \sigma)$ with $T = (V, E)$ are:

Ia. INTERIOR: Given an arrow $e \in E$ between interior nodes, contract e and set

$$\tau(v^*) = \tau(s(e)) + \tau(t(e)) - (0, (1, 0, 0), (1, 0, 0)).$$

Ib. BOUNDARY-OBSTRUCTED: Given a boundary node $v \in V$ with

$$s_1(v) = |\{e \in E \,|\, t(e) = v,\ \mu(s(e)) = 0\}|,$$
$$u_2(v) = |\{e \in E \,|\, s(e) = v,\ \mu(t(e)) = 0\}|,$$

contract $\mathrm{Star}^o(v)$ and set

$$\tau(v^*) = \sum_{v' \in \mathrm{Nbhd}^o(v)} \tau(v') - \tau(v) + (1, (0, 0, 0), (0, 0, 0)).$$

II. BOUNDARY: Given an arrow $e \in E$ between boundary nodes, contract e and set

$$\tau(v^*) = \begin{cases} \tau(s(e)) + \tau(t(e)) - (1, (0, 1, 0), (0, 1, 0)) & \text{if } \sigma(e) = u, \\ \tau(s(e)) + \tau(t(e)) - (1, (0, 0, 1), (0, 0, 1)) & \text{if } \sigma(e) = s. \end{cases}$$

We denote the resulting abstract paths by $q^{Ia}(\tau, e)$, $q^{Ib}(\tau, v)$, and $q^{II}(\tau, e)$, respectively.

Definition 3.6. A *marker* on τ is a node or arrow of the underlying tree T. We will use X as shorthand for the marker set $V \cup E$. Note that at most one of $q^{Ia}(\tau, x)$, $q^{Ib}(\tau, x)$, and $q^{II}(\tau, x)$ is defined for each $x \in X$. If one is indeed defined, we say the marker is *valid* and denote the resulting glued abstract path by $\tilde{q}(\tau, x)$.

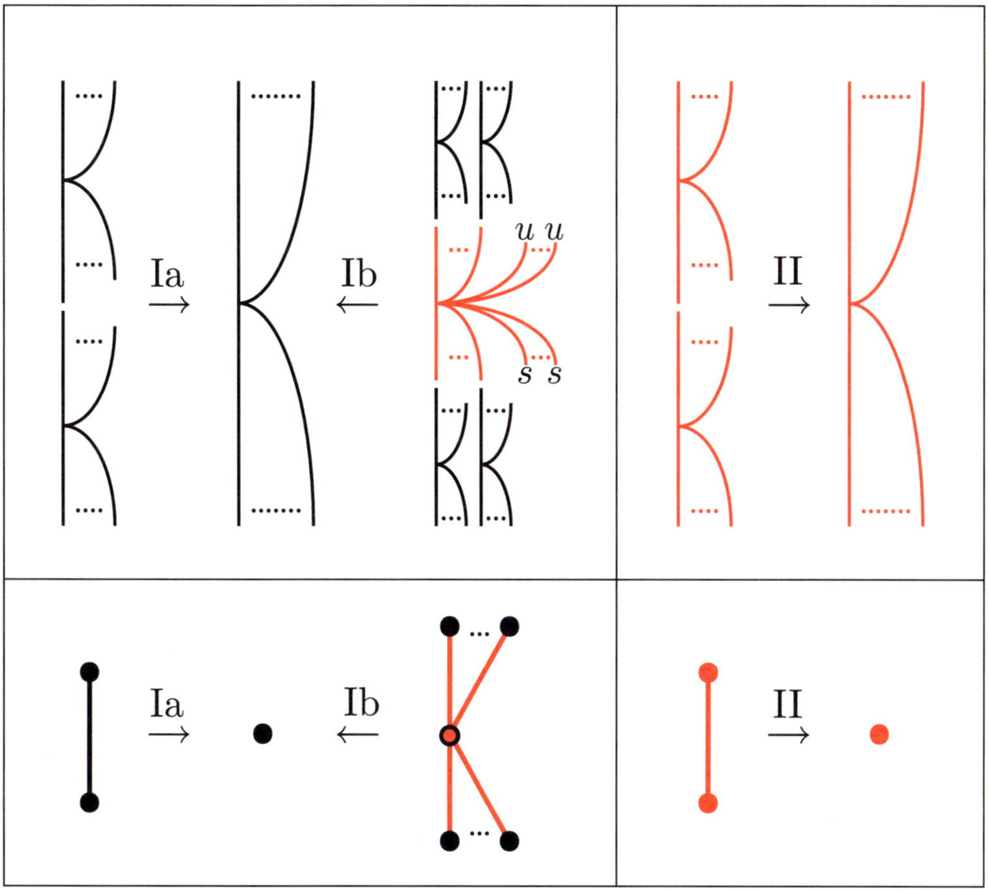

FIGURE 7. The three gluing moves are Ia (interior), Ib (boundary-obstructed), and II (boundary). Move Ia glues two adjacent interior edges to obtain a single interior edge. Move Ib is applicable in the presence of a boundary edge such that the subpath obtained by appending adjacent interior edges is legal. Move Ib replaces this subpath with a single interior edge. Move II glues two adjacent boundary edges to obtain a single boundary edge. The effect of each gluing move is to contract the underlying tree, as shown in the second row. Interior (resp., boundary) breaks and abstract edges above correspond to black (resp., red) arrows and nodes below. A red node with a black border is valid for Move Ib (a red node without a black border may or may not be valid for Move Ib).

3.2. Abstract paths of low weight

We now define a notion of weight for abstract paths. Since all vertices, edges, and paths in this section are abstract, we will drop the descriptor.

Definition 3.7. The *weight* of an edge $\varepsilon = (\mu, (o_1, u_1, s_1), (o_2, u_2, s_2))$ is given by:

$$w(\varepsilon) = \begin{cases} 1 & \text{if } \mu = 0, \\ 2 - s_1 - u_2 & \text{if } \mu = 1. \end{cases}$$

The *weight* of a path is the sum of the weights of its edges:

$$w(\tau) = \sum_{v \in V} w(\tau(v)).$$

Lemma 3.1. *Each gluing move reduces weight by 1.*

Proof. For Move Ia, the result is immediate.

For Move Ib, it suffices to show that the subpath τ' corresponding to $\mathrm{Star}^o(v)$ has weight 2, since this subpath is being replaced by an edge of weight 1. We induct on the length $l(\tau')$. If $l(\tau') = 1$, then τ' consists of a single legal boundary edge $\varepsilon = \tau(v)$, which has weight 2 by definition. If $l(\tau') > 1$, then τ' includes an interior edge ε'. We suppose that ε' precedes $\varepsilon = \tau(v)$. Then the break e between ε' and ε is stable. Let τ'' be the subpath of τ' that results from pruning ε' from τ'. Then

$$w(\tau') = w(\tau'') + 1.$$

The subpath τ'' is illegal due to a single incoming stable end. Converting this stable end to an unstable end yields a legal path τ''' with

$$w(\tau''') = w(\tau'') + 1.$$

Furthermore, τ''' is a candidate for Move II of length $l(\tau') - 1$, so by induction we have

$$w(\tau''') = 2.$$

Together these equations imply $w(\tau') = 2$. The proof is similar if ε' follows ε.

For Move II, let $\varepsilon_1 = \tau(s(e))$ and $\varepsilon_2 = \tau(t(e))$ be the consecutive boundary edges and let ε be the resulting glued boundary edge. If e is stable, then

$$\begin{aligned} w(\varepsilon_1) + w(\varepsilon_2) &= (2 - s_1(\varepsilon_1) - u_2(\varepsilon_1)) + (2 - s_1(\varepsilon_2) - u_2(\varepsilon_2)) \\ &= 1 + 2 - (s_1(\varepsilon_1) + s_1(\varepsilon_2) - 1) - (u_2(\varepsilon_1) + u_2(\varepsilon_2)) \\ &= 1 + 2 - s_1(\varepsilon) - u_2(\varepsilon) \\ &= 1 + w(\varepsilon). \end{aligned}$$

The proof is similar if e is unstable. □

In order to classify paths of low weight, we introduce several counting functions.

Definition 3.8. Given a path τ, let $O(\tau)$, $U(\tau)$, and $S(\tau)$ denote the number of breaks of each type:

$$O(\tau) = |\sigma^{-1}(o)|$$
$$U(\tau) = |\sigma^{-1}(u)|$$
$$S(\tau) = |\sigma^{-1}(s)|$$

Let $B(\tau)$ denote the number of boundary edges:

$$B(\tau) = |\mu^{-1}(1)|$$

Proposition 3.2. *For any legal path τ, we have*

$$w(\tau) = 1 + O(\tau) + B(\tau). \tag{10}$$

In particular, a legal path has weight 1 *if and only if it consists of a single interior edge.*

Proof. We induct on the length $l(\tau)$. The base case $l(\tau) = 1$ follows from the fact that a legal boundary edge has weight 2.

We now suppose that $l(\tau) > 1$. Note that Move Ib, applied to a legal boundary edge, reduces both w and B by 1 while preserving O (and l), and therefore preserves (10). So we may assume without loss of generality that τ has no legal boundary edges. Now τ falls into at least one of the following cases:

Case 1: τ has two adjacent interior edges.

Case 2: τ has a boundary edge without adjacent boundary edges.

Case 3: τ has two adjacent boundary edges.

In Case 1, we can apply Move Ia. This reduces l and O by 1 and preserves B. In Case 2, since τ is legal, we can apply Move Ib to the interior star of this (illegal) boundary edge to reduce l by at least 1. This preserves O and reduces B by 1. In Case 3, we can apply Move II. This reduces l and B by 1 and preserves O. So in all three cases, the right-hand side of (10) is reduced by 1. By Lemma 3.1, the left-hand side of (10) is reduced by 1 as well. Thus (10) is preserved. Since length is reduced in all cases, we are done by induction.

We now turn to the final claim. If τ is legal and $w(\tau) = 1$, then (10) implies that τ has no boundary edges and no interior breaks. Since breaks between adjacent interior edges are necessarily interior, we conclude that τ consists of a single interior edge. □

Corollary 3.3 (Classification of legal paths of low weight).
(i) A legal path has weight 2 if and only if it has the form Ia or Ib.
(ii) A boundary path has length 2 if and only if it has the form II.

Proof. For (i), suppose τ is legal and $w(\tau) = 2$. By (10), there are two possibilities:

Case 1: $O(\tau) = 1$ and $B(\tau) = 0$.

Case 2: $O(\tau) = 0$ and $B(\tau) = 1$.

In Case 1, τ consists of two interior edges, giving the form Ia.

In Case 2, no two interior edges may be adjacent, so τ consists of a single boundary edge $\tau(v)$ and some number of adjacent interior edges. Thus τ coincides with $\mathrm{Star}^o(v)$. Since τ is legal, it has the form Ib.

Claim (ii) is tautological. □

Corollary 3.4. *For any path τ, we have*

$$w(\tau) = 1 + O(\tau) + B(\tau) - s_1(\tau) - u_2(\tau). \tag{11}$$

Proof. We build a legal path τ' from the path τ by concatenating the abstract edge $(1, (0, 1, 0), (0, 0, 1))$ before each incoming stable end of τ and after each outgoing unstable end of τ. This adds $s_1(\tau) + u_2(\tau)$ boundary edges of weight 2, so

$$\begin{aligned} w(\tau) &= w(\tau') - 2(s_1(\tau) + u_2(\tau)) \\ &= (1 + O(\tau') + B(\tau')) - 2(s_1(\tau) + u_2(\tau)) \\ &= 1 + O(\tau) + (B(\tau) + s_1(\tau) + u_2(\tau)) - 2(s_1(\tau) + u_2(\tau)) \\ &= 1 + O(\tau) + B(\tau) - s_1(\tau) - u_2(\tau) \end{aligned}$$

where the second equality uses Proposition 3.2. □

3.2.1. Good tree-like paths

Definition 3.9. A path is *tree-like* if every edge has exactly one outgoing end. For such a path τ, let $B_s(\tau)$ denote the number of boundary edges with stable outgoing end:

$$B_s(\tau) = |\{v \in V \mid \mu(v) = 1 \text{ and } s_2(v) = 1\}|.$$

Definition 3.10. A path is *good* if each end is interior or unstable. A break is *good* if it is interior or unstable. Let $G(\tau)$ denote the number of good breaks in τ:

$$G(\tau) = O(\tau) + U(\tau).$$

We will use the following corollary of Corollary 3.4 to characterize good tree-like paths of low weight.

Corollary 3.5. *If τ is a legal tree-like path or a good tree-like path, then*

$$w(\tau) = 1 + G(\tau) + B_s(\tau). \tag{12}$$

Proof. If τ ends in o or s, then

$$\begin{aligned} w(\tau) &= 1 + O(\tau) + B(\tau) \\ &= 1 + O(\tau) + (U(\tau) + B_s(\tau)) \\ &= 1 + (O(\tau) + U(\tau)) + B_s(\tau) \\ &= 1 + G(\tau) + B_s(\tau). \end{aligned}$$

If τ ends in u, then

$$\begin{aligned} w(\tau) &= 1 + O(\tau) + B(\tau) - 1 \\ &= 1 + O(\tau) + (U(\tau) + 1 + B_s(\tau)) - 1 \\ &= 1 + (O(\tau) + U(\tau)) + B_s(\tau) \\ &= 1 + G(\tau) + B_s(\tau). \end{aligned}$$

□

Proposition 3.6 (Classification of good tree-like paths of low weight).
(i) A good tree-like path has weight 1 if and only if it has a form in (a) of Figure 8.
(ii) A good tree-like path has weight 2 if and only if it has a form in (c) or (d) of Figure 8.

Proof. Let τ be a good tree-like path, with final edge ε.

For (i), if τ has weight 1 then by (12) all breaks of τ must be stable and follow interior edges. So if ε is interior, then no edges may precede ε. And if ε is boundary, then each incoming end of ε is preceded by at most one interior edge. These two cases yield the two forms in (a).

For (ii), if τ has weight 2 then by (12) all breaks but one are both stable and follow interior edges. Let e be this rogue break. Note that if ε is interior than it is necessarily preceded by an interior or unstable break. We are therefore left with six mutually exclusive cases:

Case 1: e is interior and ε is interior.
Case 2: e is interior and ε is boundary.
Case 3: e is unstable and ε is interior.
Case 4: e is unstable, ε is boundary, and $t(e)$ is interior.
Case 5: e is unstable, ε is boundary, and $t(e)$ is boundary.
Case 6: e is stable and ε is boundary.

Note that in Cases 3 and 4, the edge ε' preceding the unstable break e is boundary and all edges adjacent to ε' are interior. Since τ is good, the interior star of ε' is therefore legal.

In each case, we claim that τ must have a particular form. The six forms are shown sequentially from left to right in the second row of Figure 8. In each case, the line of argument is to perform a gluing move on τ, apply (i) to force the form of the resulting good tree-like path of weight 1, and then reverse the gluing move to recover the form of τ. In Cases 1 and 2, we apply Move Ia at e. In Cases 3 and 4, we apply Move Ib at the edge ε' preceding e. In Cases 5 and 6, we apply Move II at e. The six resulting paths are shown sequentially in the third row of Figure 8 (the seventh path does not play a role in this argument). Reversing the gluing move, we obtain the six forms in the second row. □

The following corollary of Proposition 3.6 is crucial for generalizing the structure equation to bigraphs.

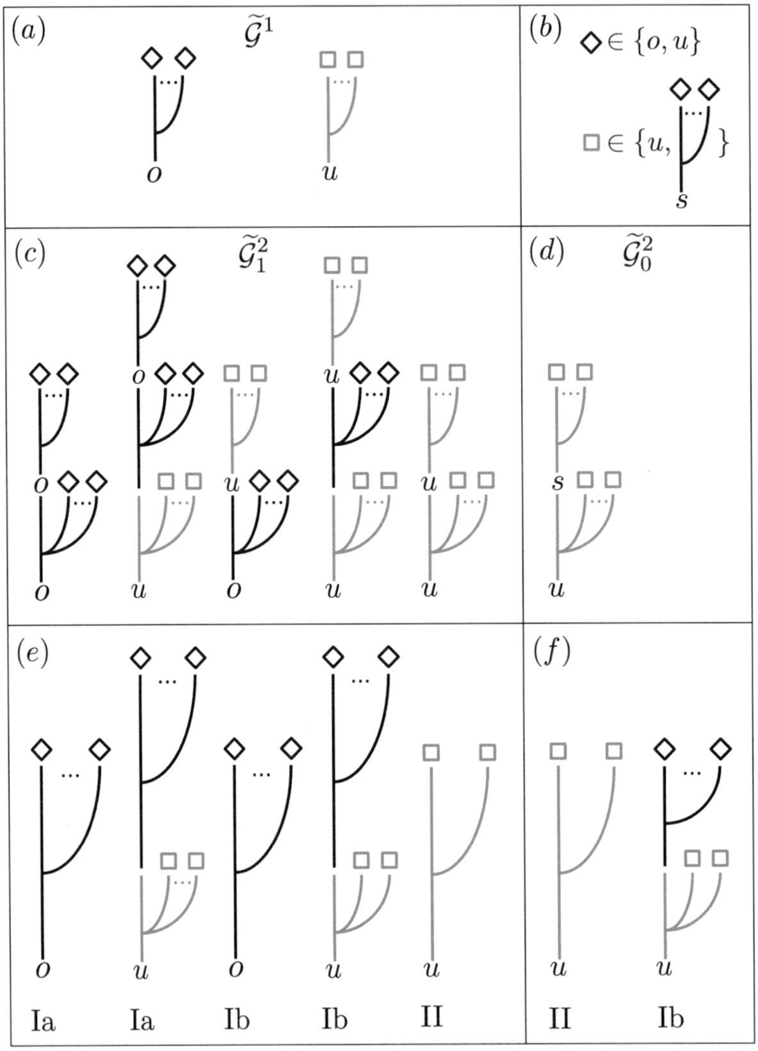

FIGURE 8. The forms of a good tree-like weight 1 path are shown in (a), where the diamond and square symbols are members of the sets in (b) and ellipses indicate any finite number of incoming ends (including zero). The forms of a good, tree-like weight 2 path are shown in (c) and (d). Those with one good break are in (c), each admitting a unique gluing shown directly below in (e). Those with no good breaks are shown in (d), each admitting the two gluings shown in (f). The notation $\widetilde{\mathcal{G}}^k_m$ is introduced in Section 4.3.

Corollary 3.7 (The key to good breaks)**.** *Let τ be a good tree-like path of weight 2. Then τ has at most one good break. Furthermore:*
(i) If τ has one good break, then τ can be glued in precisely one way.
(ii) If τ has no good breaks, then τ can be glued in precisely two ways.

Proof. The path τ has at most one good break by (12). By Proposition 3.6, τ has one of the six forms shown in (c) and (d) of Figure 8. Each form in (c) has a good break and may be glued in precisely the one way shown directly below in (e). The form in (d) has no good break and may be glued in precisely the two ways shown in (f). □

4. Bigraphs

We now introduce the bigraph $\widetilde{\mathcal{H}}$ associated to a higraph $\mathcal{H}$. The "b" stands for blow-up or boundary and "bi" signifies directedness. The bigraph $\widetilde{\mathcal{H}} = (\widetilde{\mathcal{V}}, \widetilde{\mathcal{E}})$ covers the higraph $\mathcal{H} = (\mathcal{V}, \mathcal{E})$ in the sense that there is a 3-to-1 map $\eta : \widetilde{\mathcal{V}} \to \mathcal{V}$ which induces a map on edges that preserves the size of source and target. The three lifts of a vertex in $\mathcal{V}$ are labeled by the three elements of the abstract vertex set $\mathcal{V}_{ab}$, and are called interior, unstable, and stable. The $2^{|I|+|J|+1}$ lifts of an edge $(I, J) \in \mathcal{E}$ are precisely the compatible abstract edges, of which $2^{|I|+|J|}$ are interior and $2^{|I|+|J|}$ are boundary. We make this precise in the following definitions, which are followed by examples in Figures 9 and 10.

Definition 4.1. The *bigraph* $\widetilde{\mathcal{H}} = (\widetilde{\mathcal{V}}, \widetilde{\mathcal{E}})$ associated to the higraph $\mathcal{H}$ consists of a vertex set

$$\widetilde{\mathcal{V}} = \mathcal{V} \times \{o, u, s\}$$

and an edge set

$$\widetilde{\mathcal{E}} \subset \mathbb{F} \times 2^{\widetilde{\mathcal{V}}} \times 2^{\widetilde{\mathcal{V}}}$$

with $(\mu, \tilde{I}, \tilde{J}) \in \widetilde{\mathcal{E}}$ if and only if $|\eta(\tilde{I})| = |\tilde{I}|$, $|\eta(\tilde{J})| = |\tilde{J}|$, $(\eta(\tilde{I}), \eta(\tilde{J}))$ in $\mathcal{E}$ and $g(\tilde{I}, \tilde{J}, \mu)$ in $\mathcal{E}_{ab}$.

Here $\eta : 2^{\widetilde{\mathcal{V}}} \to 2^{\mathcal{V}}$ is induced by the projection $\eta : \widetilde{\mathcal{V}} \to \mathcal{V}$ and

$$g : \mathbb{F} \times 2^{\widetilde{\mathcal{V}}} \times 2^{\widetilde{\mathcal{V}}} \to \mathbb{F} \times (\mathbb{N} \times \mathbb{N} \times \mathbb{N}) \times (\mathbb{N} \times \mathbb{N} \times \mathbb{N})$$

is defined by

$$g(\mu, \tilde{I}, \tilde{J}) = (\mu, (\tilde{o}(\tilde{I}), \tilde{u}(\tilde{I}), \tilde{s}(\tilde{I})), (\tilde{o}(\tilde{J}), \tilde{u}(\tilde{J}), \tilde{s}(\tilde{J})))$$

where $\tilde{o}(\tilde{I}) = |\{(K, o) \in \tilde{I}\}|$, $\tilde{u}(\tilde{I}) = |\{(K, u) \in \tilde{I}\}|$, and $\tilde{s}(\tilde{I}) = |\{(K, s) \in \tilde{I}\}|$.

Definition 4.2. The *boundary bigraph* $\overline{\mathcal{H}} = (\overline{\mathcal{V}}, \overline{\mathcal{E}})$ associated to the higraph $\mathcal{H}$ is the directed hypergraph given by the boundary vertices and boundary edges of $\widetilde{\mathcal{H}}$. Equivalently, $\overline{\mathcal{H}}$ has the vertex set

$$\overline{\mathcal{V}} = \mathcal{V} \times \{u, s\}$$

and the edge set

$$\overline{\mathcal{E}} \subset 2^{\overline{\mathcal{V}}} \times 2^{\overline{\mathcal{V}}}$$

with $(\bar{I}, \bar{J}) \in \overline{\mathcal{E}}$ if and only if $|\eta(\bar{I})| = |\bar{I}|$, $|\eta(\bar{J})| = |\bar{J}|$, and $(\eta(\bar{I}), \eta(\bar{J})) \in \mathcal{E}$.

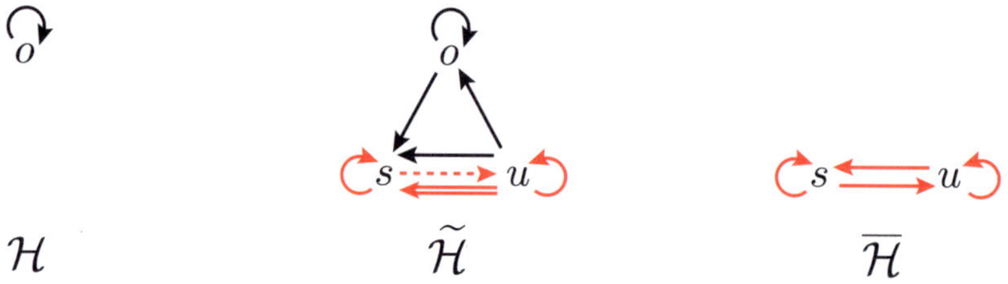

FIGURE 9. The higraph $\mathcal{H}$ above consists of a single vertex and a single loop. The loop lifts to four interior edges and four boundary edges in the bigraph $\widetilde{\mathcal{H}}$. The dotted edge has weight 0, the doubled edge has weight 2, and all other edges have weight 1. Only the boundary vertices and edges are present in the boundary bigraph $\overline{\mathcal{H}}$.

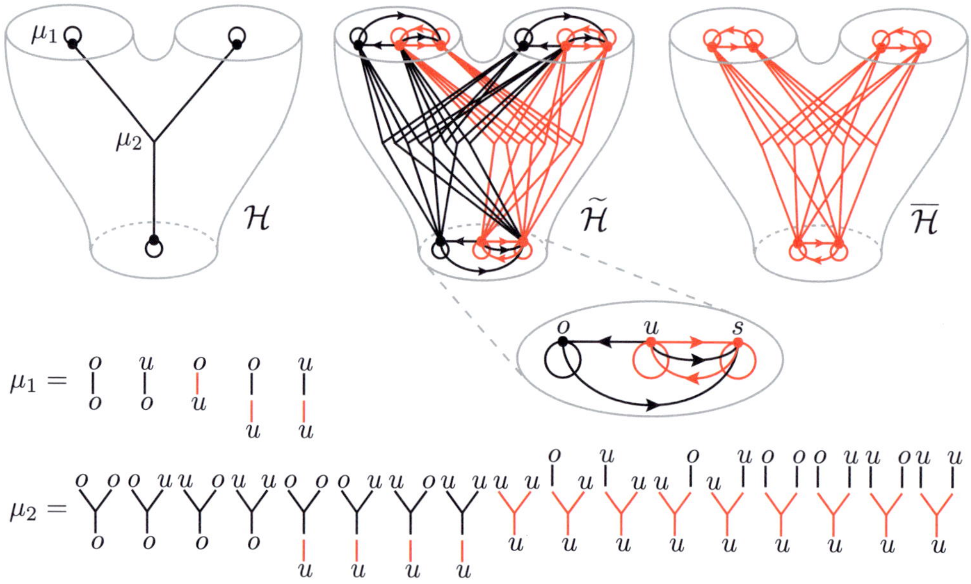

FIGURE 10. The higraph $\mathcal{H}$ above consists of three vertices and four edges. Three edges are loops and the remaining edge ε has two vertices as source and one vertex as target. The edge ε lifts to eight interior edges and eight boundary edges in the bigraph $\widetilde{\mathcal{H}}$. Only the boundary vertices and edges are present in the boundary bigraph $\overline{\mathcal{H}}$. Edge weights are not indicated.

4.1. Paths

Note that $\widetilde{\mathcal{H}}$ and $\overline{\mathcal{H}}$ are not higraphs because there are non-loop cycles in $\widetilde{\mathcal{H}}$ and $\overline{\mathcal{H}}$, though the edges of a cycle project to a common loop in $\mathcal{H}$. Furthermore, there are two lifts of each edge $(I, J) \in \mathcal{E}$ to $\widetilde{\mathcal{E}}$ with the same source and target, namely $(I \times \{u\}, J \times \{s\}, 0)$ and $(I \times \{u\}, J \times \{s\}, 1)$. Nevertheless, we may define paths just as before.

Definition 4.3. A *path* $\tilde{\rho} = (T, \tilde{\rho})$ in a bigraph $\widetilde{\mathcal{H}}$ consists of a non-empty, directed tree $T = (V, E)$ together with a map $\tilde{\rho} : V \to \widetilde{\mathcal{E}}$ such that the following properties hold.

(1) The map $\tilde{\rho}$ induces a well-defined map $\tilde{\rho} : E \to \widetilde{\mathcal{V}}$ such that

$$\{\tilde{\rho}(e)\} = t(\tilde{\rho}(s(e))) \cap s(\tilde{\rho}(t(e))).$$

(2) If distinct arrows e, e' have the same source or the same target, then $\tilde{\rho}(e) \neq \tilde{\rho}(e')$.

Paths ρ and ρ' are *isomorphic* if there is a digraph isomorphism $f : T \to T'$ of the associated directed trees such that $\tilde{\rho} = \tilde{\rho}' f$.

As with higraphs, paths in a bigraph are unique up to unique isomorphism and we regard isomorphic paths as the same. We may similarly define the notion of a path $\bar{\rho}$ in $\overline{\mathcal{H}}$.

Let $\widetilde{\mathcal{P}}$ denote the set of paths in $\widetilde{\mathcal{H}}$, and let $\overline{\mathcal{P}}$ denote the set of paths in $\overline{\mathcal{H}}$. The operations of concatenation, gluing, and refinement carry over without change. Since each path $\tilde{\rho}$ determines an abstract path $g(\tilde{\rho})$, we may pull back the three gluing moves from abstract paths to bigraph paths.

Definition 4.4. A *k-step gluing sequence* on a path $\tilde{\rho}_1$ is a k-tuple

$$(x_1, x_2, \ldots, x_k)$$

with x_i a valid marker on $\tilde{\rho}_i$ and $\tilde{\rho}_{i+1} = \tilde{q}(\tilde{\rho}_i, x_i)$. We may use the notation

$$\tilde{\rho}_1 \xrightarrow{x_1} \tilde{\rho}_2 \xrightarrow{x_2} \cdots \xrightarrow{x_k} \tilde{\rho}_{k+1}$$

to emphasize the paths along the way.

As with the higraph case, gluing sequences in a bigraph are unique up to unique isomorphism. Unlike the higraph case, the number of k-step gluing sequences between two fixed paths in a bigraph may not be a multiple of $k!$. Indeed, Figure 11 gives an example with exactly eight 3-step gluing sequences for the bigraph in Figure 9. However, the analogue of Lemma 2.1 does hold when $k = 2$.

Lemma 4.1. *The number of 2-step gluing sequences between any two fixed paths in a bigraph is even.*

This is a special case of Proposition 4.8, whose proof we defer to Section 4.6.2.

4.2. Path DGA

Definition 4.5. The *path DGA* of the bigraph $\widetilde{\mathcal{H}}$ over the field $\mathbb{F}$ consists of a graded algebra $\widetilde{\mathcal{A}}$ equipped with a differential $\tilde{\delta} : \widetilde{\mathcal{A}} \to \widetilde{\mathcal{A}}$. The algebra $\widetilde{\mathcal{A}}$ has underlying vector space $\mathbb{F}\langle\widetilde{\mathcal{P}}\rangle$, with the subspace $\widetilde{\mathcal{A}}^k$ in grading k spanned by the set $\widetilde{\mathcal{P}}^k$ of paths of weight k. Multiplication is given by concatenation as before.

The value of $\tilde{\delta}$ on an interior edge $\tilde{\varepsilon}$ is the sum of all refinements of $\tilde{\varepsilon}$ to a path of weight 2. The value of $\tilde{\delta}$ on a boundary edge $\bar{\varepsilon}$ is the sum of all refinements of $\bar{\varepsilon}$ to a boundary path of length 2.

The differential $\tilde{\delta}$ has degree 1, since a boundary refinement of a boundary edge $\bar{\varepsilon}$ has length 2 if and only if it has weight $w(\bar{\varepsilon})+1$ by (11). We may define $\tilde{\delta}$ directly on a path $\tilde{\rho} = ((V, E), \tilde{\rho})$ by

$$\tilde{\delta}\tilde{\rho} = \sum_{v \in V^0} \sum_{\tilde{\rho}' \in \widetilde{\mathcal{P}}^2} r((\tilde{\rho}, v), \tilde{\rho}') + \sum_{v \in V^1} \sum_{\bar{\rho}' \in \overline{\mathcal{P}}^2} r((\tilde{\rho}, v), \bar{\rho}'). \tag{13}$$

We now define the adjoint map $\tilde{\partial}$ with respect to the symmetric, bilinear pairing

$$\langle \cdot, \cdot \rangle : \widetilde{\mathcal{A}} \times \widetilde{\mathcal{A}} \to \mathbb{F}$$

with $\langle \tilde{\rho}_1, \tilde{\rho}_2 \rangle = 1$ if and only if $\tilde{\rho}_1 = \tilde{\rho}_2$.

Definition 4.6. The map $\tilde{\partial} : \widetilde{\mathcal{A}} \to \widetilde{\mathcal{A}}$ is defined on the path $\tilde{\rho}$ with underlying tree $T = (V, E)$ by

$$\tilde{\partial}\tilde{\rho} = \sum_{x \in X} \tilde{q}(\tilde{\rho}, x) \tag{14}$$

where $\tilde{q}(\tau, x)$ is defined to be zero on invalid input.

Theorem 4.2. *We have*

$$\langle \tilde{\partial} a, b \rangle = \langle a, \tilde{\delta} b \rangle$$

for all $a, b \in \widetilde{\mathcal{A}}$. *Furthermore,* $\tilde{\partial}^2 = 0$ *and* $\tilde{\delta}^2 = 0$.

Proof. For the first statement, by linearity it suffices to verify that

$$\langle \tilde{\partial}\tilde{\rho}_1, \tilde{\rho}_2 \rangle = \langle \tilde{\rho}_1, \tilde{\delta}\tilde{\rho}_2 \rangle$$

for all paths $\tilde{\rho}_1$ and $\tilde{\rho}_2$. The left-hand side counts all 1-step gluing sequences $\tilde{\rho}_1 \xrightarrow{x_1} \tilde{\rho}_2$. The right-hand side counts all ways to refine an interior edge of $\tilde{\rho}_2$ to a path of weight 2 to obtain $\tilde{\rho}_1$, plus all ways to refine a boundary edge of $\tilde{\rho}_2$ to a boundary path of length 2 to obtain $\tilde{\rho}_1$. These counts coincide by Corollary 3.3.

To prove $\tilde{\partial}^2 = 0$, it suffices to show that $\langle \tilde{\partial}^2 \tilde{\rho}_1, \tilde{\rho}_3 \rangle$ vanishes for all paths $\tilde{\rho}_1$ and $\tilde{\rho}_3$. Indeed, the term $\langle \tilde{\partial}^2 \tilde{\rho}_1, \tilde{\rho}_3 \rangle$ counts the number of 2-step gluing sequences from $\tilde{\rho}_1$ to $\tilde{\rho}_3$, which is even by Lemma 4.1. Now $\tilde{\delta}^2 = 0$ by adjointness as before. ∎

4.3. Tree-like higraphs and bigraphs

The following notions for abstract paths readily carry over to bigraphs as well.

Definition 4.7. A higraph or bigraph is *tree-like* if the target of each edge consists of exactly one vertex. A path in a bigraph is *good* if each end is interior or unstable. A break is *good* if it is interior or unstable.

Clearly the bigraph $\widetilde{\mathcal{H}}$ is tree-like if and only if the underlying higraph $\mathcal{H}$ is tree-like. Let $\widetilde{\mathcal{G}}^k$ be the set of good paths in $\widetilde{\mathcal{H}}$ of weight k. Let $\widetilde{\mathcal{G}}^k_m$ be the set of good paths in $\widetilde{\mathcal{H}}$ of weight k with m good breaks. Note that $\widetilde{\mathcal{G}}^1 = \widetilde{\mathcal{G}}^1_0$ and $\widetilde{\mathcal{G}}^2 = \widetilde{\mathcal{G}}^2_1 \cup \widetilde{\mathcal{G}}^2_0$ by (12); the forms of these paths are shown in Figure 8.

Definition 4.8. We define the element $\tilde{D} \in \widetilde{\mathcal{A}}$ by

$$\tilde{D} = \sum_{\tilde{\rho} \in \widetilde{\mathcal{G}}^1} \tilde{\rho}.$$

Theorem 4.3. *If $\mathcal{H}$ is tree-like, then the element $\tilde{D}$ satisfies the structure equation:*

$$\tilde{\delta}\tilde{D} = \tilde{D} \circ \tilde{D}.$$

Proof. We claim

$$\tilde{\delta}\tilde{D} = \sum_{\tilde{\rho} \in \widetilde{\mathcal{G}}^2_1} \tilde{\rho} + \sum_{\tilde{\rho} \in \widetilde{\mathcal{G}}^2_0} 2\tilde{\rho} = \sum_{\tilde{\rho} \in \widetilde{\mathcal{G}}^2_1} \tilde{\rho} = \tilde{D} \circ \tilde{D}. \tag{15}$$

The left-hand equality holds by Corollary 3.7. The middle equality holds over $\mathbb{F}$. The right-hand equality holds because each good path of weight 2 with one good break arises as a product of two (not necessarily distinct) good paths of weight 1 in a unique way (as specified by cutting at the good break). ☐

4.4. Balanced higraphs and bigraphs

The reader may check that Theorem 4.3 is always false if $\mathcal{H}$ is not tree-like. On the other hand, there is a duality at play, which switches the roles of unstable and stable vertices. Dual to tree-like higraphs are *rake-like* higraphs, where the source of each edge consists of a single vertex. If $\mathcal{H}$ is rake-like, and we re-define good breaks to be interior or stable ones, then Theorem 4.3 continues to hold. More generally, we can independently define the notion of good at each vertex of $\mathcal{H}$, provided each edge of $\mathcal{H}$ has a compatible shape.

Definition 4.9. A *vertex assignment* on a higraph $\mathcal{H} = (\mathcal{V}, \mathcal{E})$ is a map $\psi : \mathcal{V} \to \{1, 2\}$ such that

$$|s(\varepsilon) \cap \psi^{-1}(2)| + |t(\varepsilon) \cap \psi^{-1}(1)| = 1 \tag{16}$$

for every edge $\varepsilon \in \mathcal{E}$. In this case, we say that $\mathcal{H}$ is *balanced* by ψ. The set of ψ-good vertices of $\widetilde{\mathcal{H}}$ is given by

$$\widetilde{\mathcal{V}}_\psi = (\psi^{-1}(1) \times \{o, u\}) \cup (\psi^{-1}(2) \times \{o, s\}) \subset \widetilde{\mathcal{V}}.$$

A path $\tilde{\rho}$ is ψ-good if each end is ψ-good. A break e of $\tilde{\rho}$ is ψ-good if $\tilde{\rho}(e)$ is ψ-good. Let $\widetilde{\mathcal{P}}^k(\psi)$ denote the set of ψ-good paths of weight k. Let $\widetilde{\mathcal{P}}^k_m(\psi)$ denote the set of ψ-good paths of weight k with m ψ-good breaks.

Note if $\psi \equiv 1$ then $\widetilde{\mathcal{P}}^k_m(\psi) = \widetilde{\mathcal{G}}^k_m$, and ψ balances $\mathcal{H}$ if and only if $\mathcal{H}$ is tree-like. If $\psi \equiv 2$, then ψ-good breaks are interior or stable ones, and ψ balances $\mathcal{H}$ if and only if $\mathcal{H}$ is rake-like.

Remark 4.1. The proof of the monopole surgery exact triangle is modeled by a higraph with $\psi = 1$ for three vertices and $\psi = 2$ for four vertices, as explained in Figure 19.

Definition 4.10. We define the element $\tilde{D}_\psi \in \widetilde{\mathcal{A}}$ by

$$\tilde{D}_\psi = \sum_{\tilde{\rho} \in \widetilde{\mathcal{P}}^1(\psi)} \tilde{\rho}.$$

Corollary 4.4. *If $\mathcal{H}$ is balanced by ψ, then the element $\tilde{D}_\psi$ satisfies the following structure equation:*

$$\tilde{\delta}\tilde{D}_\psi = \tilde{D}_\psi \circ \tilde{D}_\psi.$$

Proof. Given $\mathcal{H} = (\mathcal{V}, \mathcal{E})$ balanced by ψ, we set $\mathcal{H}' = (\mathcal{V}, \mathcal{E}')$ with $\mathcal{E}$ in bijection with $\mathcal{E}'$ via

$$\varepsilon = (I, J) \mapsto \varepsilon' = \left(\left(I \cap \psi^{-1}(1)\right) \cup \left(J \cap \psi^{-1}(2)\right), \left(J \cap \psi^{-1}(1)\right) \cup \left(I \cap \psi^{-1}(2)\right) \right).$$

In other words, we flip each $\psi = 2$ end of each edge from source to target and vice versa. This bijection establishes a weight-preserving correspondence between ψ-good paths in $\widetilde{\mathcal{H}}$ and good paths in $\widetilde{\mathcal{H}}'$, with s and u exchanging roles at $\psi = 2$ vertices. So the equation

$$\tilde{\delta}\tilde{D}_\psi = \sum_{\tilde{\rho} \in \widetilde{\mathcal{P}}^2_1(\psi)} \tilde{\rho} + \sum_{\tilde{\rho} \in \widetilde{\mathcal{P}}^2_0(\psi)} 2\tilde{\rho} = \sum_{\tilde{\rho} \in \widetilde{\mathcal{P}}^2_1(\psi)} \tilde{\rho} = \tilde{D}_\psi \circ \tilde{D}_\psi$$

on $\widetilde{\mathcal{H}}$ is equivalent to equation (15) on $\widetilde{\mathcal{H}}'$. The latter holds since $\mathcal{H}'$ is tree-like by equation (16). □

4.5. Boundary bigraphs

We briefly return to boundary bigraphs.

Definition 4.11. The *path DGA* of the boundary bigraph $\overline{\mathcal{H}}$ over the field $\mathbb{F}$ consists of a graded algebra $\overline{\mathcal{A}}$ equipped with a differential $\bar{\delta} : \overline{\mathcal{A}} \to \overline{\mathcal{A}}$. The algebra $\overline{\mathcal{A}}$ has underlying vector space $\mathbb{F}\langle\overline{\mathcal{P}}\rangle$, with the subspace $\overline{\mathcal{A}}^k$ in grading k spanned by the set $\overline{\mathcal{P}}^k$ of paths of length k. Multiplication is given by concatenation.

The value of $\bar{\delta}$ on a boundary edge $\bar{\varepsilon}$ equals the sum of all refinements of $\bar{\varepsilon}$ to a boundary path of length 2. In particular, $\overline{\mathcal{A}}$ sits inside $\widetilde{\mathcal{A}}$ as a differential algebra.

The arguments in Section 2 for higraphs carry over essentially without change to boundary bigraphs. In particular, $\bar{\delta}$ is indeed a differential and the element

$$\bar{D} = \sum_{\bar{\rho}\in\overline{\mathcal{P}}^1} \bar{\rho}$$

of $\overline{\mathcal{A}}$ satisfies the structure equation

$$\bar{\delta}\bar{D} = \bar{D} \circ \bar{D}$$

without restriction on the shape of edges.

4.6. Gluing sequences

We now take a closer look at gluing sequences in order to prove Lemma 2.1 and Lemma 4.1. A path in a higraph with two breaks has a 2-step gluing sequence. However, there exist paths in a bigraph, such as path (d) in Figure 8, with two available gluing moves but no 2-step gluing sequence. The subtlety is that the breaks in a gluing sequence live on different paths. At least for bigraphs, it is not immediately clear how to go about reordering them. The key will be to pull the breaks or markers in a gluing sequence up to breaks on the initial path. Though the higraph case is straightforward, we consider it in detail as a model for the bigraph case.

4.6.1. Higraphs

Let ρ_1 be a path in a higraph $\mathcal{H}$. We will define a bijection between k-tuples of distinct breaks on ρ_1 and k-step gluing sequences on ρ_1. We relate these notions by populating a lower triangular matrix A whose (i,j)-entry e^i_j is a break on ρ_j, with $\rho_j = q(\rho_{j-1}, e^{j-1}_{j-1})$. Here we use the bijection π on arrows as defined following (8).

Let $(e^1_1, e^2_1, \dots e^k_1)$ be a k-tuple of distinct breaks on a path ρ_1, thought of as the first column of A. We populate A one column at a time from left to right as follows. For each j from 2 to k, we set

$$\rho_j = q(\rho_{j-1}, e^{j-1}_{j-1})$$

and

$$e^i_j = \pi(e^i_{j-1})$$

for $j \le i \le k$. The diagonal of A is a k-step gluing sequence $(e^1_1, e^2_2, \dots, e^k_k)$ on ρ_1.

Let $(e^1_1, e^2_2, \dots e^k_k)$ be a k-step gluing sequence on ρ_1, thought of as the diagonal of A. We populate A one column at a time from right to left as follows. For each j from k down to 2, we have

$$\rho_j = q(\rho_{j-1}, e^{j-1}_{j-1})$$

and can thus set

$$e^i_{j-1} = \pi^{-1}(e^i_j)$$

for $j \le i \le k$. The first column of A is a k-tuple $(e^1_1, e^2_1, \dots e^k_1)$ of distinct breaks on ρ_1.

Proposition 4.5. *Let ρ_1 be a path in a higraph. The map*

$$(e_1^1, e_1^2, \ldots e_1^k) \mapsto (e_1^1, e_2^2, \ldots e_k^k)$$

gives a bijection between k-tuples of distinct breaks on ρ_1 and k-step gluing sequences on ρ_1. Furthermore, if two k-tuples differ by a permutation, then the corresponding k-step gluing sequences terminate in the same path ρ_{k+1}.

Proof. The map is a bijection as we have also defined its inverse above.

For the latter statement, consider two k-tuples $(e_1^1, e_1^2, e_1^3, \ldots e_1^k)$ and $(f_1^1, f_1^2, f_1^3, \ldots, f_1^k)$ of distinct breaks on ρ_1 which differ by a permutation, which we may assume is a transposition of neighbors. We may reduce further to the case of the transposition of the first two entries by induction on k, since if $e_1^1 = f_1^1$ then $(e_2^2, \ldots, e_2^k)$ and $(f_2^2, \ldots, f_2^k)$ are $(k-1)$-tuples of distinct breaks on $\rho_2 = q(\rho_1, e_1^1) = q(\rho_1, f_1^1)$. Note that the terminal path of $(e_2^2, \ldots, e_2^k)$ is the same as that of $(e_1^1, e_1^2, e_1^3, \ldots e_1^k)$ by definition, and similarly for $(f_2^2, \ldots, f_2^k)$.

So we suppose $e_1^1 = f_2^1$, $e_2^1 = f_1^1$, and $e_1^i = f_1^i$ for $3 \leq i \leq k$. Let T_3^e and T_3^f be the trees underlying the paths $q(q(\rho_1, e_1^1), e_2^2))$ and $q(q(\rho_1, f_1^1), f_2^2))$. When e_1^1 and e_1^2 are adjacent, these trees result from contracting T_1 along the same two-arrow subtree T', and we therefore have an isomorphism $f : T_3^e \to T_3^f$ which identifies $\pi_{e_2^2}(\pi_{e_1^1}(v))$ with $\pi_{f_2^2}(\pi_{f_1^1}(v))$ for each $v \in V_1$, and identifies the arrow e_3^i with f_3^i for $3 \leq i \leq k$. Furthermore, the map f is an isomorphism of paths, since for both paths the edge over the node v^* to which T' contracts has the same source and target as the subpath of ρ_1 supported by T'. Finally, if e_1^1 and e_1^2 are not adjacent, then the isomorphism of paths is clear.

Therefore the paths in the gluing sequences $(e_1^1, e_2^2, e_3^3, \ldots, e_k^k)$ and $(f_1^1, f_2^2, f_3^3, \ldots, f_k^k)$ are identified from ρ_3 to ρ_{k+1}. □

Proof of Lemma 2.1. Via Proposition 4.5, the set of k-step gluing sequences between any two fixed paths admits a free action of the group of permutations on k elements. □

4.6.2. Bigraphs

We now take an analogous, though more involved, approach to bigraphs. We will define a bijection between certain k-tuples of markers on ρ_1 and k-step gluing sequences on ρ_1. Recall the notion of a marker from Definition 3.6.

Definition 4.12. A marker x on $\tilde{\rho}_1$ is *valid* if $\tilde{q}(\tilde{\rho}_1, x)$ is defined. A k-tuple $(x_1^1, \ldots, x_1^k)$ of markers on $\tilde{\rho}_1$ is *valid* if its coordinates are valid and distinct.

Suppose x_1^1 is a valid marker on $\tilde{\rho}_1$, and set $\tilde{\rho}_2 = \tilde{q}(\tilde{\rho}_1, x_1^1)$. Let $T' = (V', E')$ be the subtree of T_1 that is contracted under the gluing move, and let v^* be the node in T_2 onto which T' contracts. Recall that π denotes both the map $\pi : V_1 \to V_2$ which is injective on $V_1 - V'$ with image $V_2 - \{v^*\}$, and the bijection $\pi : E_1 - E' \to E_2$. We thus have maps $\pi : X_1 - E' \to X_2$ and $\pi^{-1} : X_2 - \{v^*\} \to X_1$ which are mutual inverses whenever their composition is defined.

We define maps $\tilde{\pi} : X_1 - E' \to X_2$ and $\tilde{\pi}^{-1} : X_2 \to X_1$ as follows:

$$\tilde{\pi}(x) = \begin{cases} \pi(s(x)) & \text{if } x \text{ is a boundary arrow and } t(x) = x_1^1, & \text{(1a)} \\ \pi(t(x)) & \text{if } x \text{ is a boundary arrow and } s(x) = x_1^1, & \text{(1b)} \\ \pi(x) & \text{otherwise.} & \text{(1c)} \end{cases}$$

$$\tilde{\pi}^{-1}(y) = \begin{cases} (\pi^{-1}(y), x_1^1) & \text{if } y \text{ is a boundary node and } (y, \pi(x_1^1)) \in E_2, & \text{(2a)} \\ (x_1^1, \pi^{-1}(y)) & \text{if } y \text{ is a boundary node and } (\pi(x_1^1), y) \in E_2, & \text{(2b)} \\ x_1^1 & \text{if } y = \pi(x_1^1), & \text{(2c)} \\ t(x_1^1) & \text{if } y = \pi(t(x_1^1)) \text{ and } x_1^1 \text{ is unstable,} & \text{(2d)} \\ s(x_1^1) & \text{if } y = \pi(s(x_1^1)) \text{ and } x_1^1 \text{ is stable or interior,} & \text{(2e)} \\ \pi^{-1}(y) & \text{otherwise.} & \text{(2f)} \end{cases}$$

Note that in (1a) and (1b), x_1^1 is a marker for Move Ib. In (2a), (2b), and (2c), x_1^1 is a marker for Move Ib. In (2d), x_1^1 is a marker for Move II, while in (2e), x_1^1 is a marker for Move II or Move Ia. In (2c), (2d), and (2e), y coincides with the node v^*.

The following facts will be useful in the proof of Lemma 4.6 below:

Fact 1: The maps $\pi|_{X_1 - V' - E'}$ and $\pi^{-1}|_{X_2 - \{v^*\}}$ are mutual inverses and satisfy $\tilde{\rho}_2 \pi = \tilde{\rho}_1$ and $\tilde{\rho}_2 = \tilde{\rho}_1 \pi^{-1}$. In particular, these maps preserve abstract vertex and edge type.

Fact 2: If x_1^1 is a valid marker for Move Ib, then x_1^1 is boundary while $\pi(x_1^1)$ is interior and thus not valid.

Lemma 4.6. *The maps $\tilde{\pi} : X_1 - E' \to X_2$ and $\tilde{\pi}^{-1} : X_2 \to X_1$ are well-defined. Furthermore: (i) If $y \in X_2$, then $\tilde{\pi}(\tilde{\pi}^{-1}(y)) = y$. If y is valid, then $\tilde{\pi}^{-1}(y)$ is distinct from x_1^1 and valid.*
(ii) If $x \in X_1$ is distinct from x_1^1 and valid, then $\tilde{\pi}(x)$ is defined and $\tilde{\pi}^{-1}(\tilde{\pi}(x)) = x$.

Proof. The cases within each definition are mutually exclusive. The map $\tilde{\pi}$ is well-defined because its domain coincides with that of π. For the map $\tilde{\pi}^{-1}$, there are two issues to check. First, Fact 1 implies that $(\pi^{-1}(y), x_1^1)$ is an arrow of T_1 in (2a) and that $(x_1^1, \pi^{-1}(y))$ is an arrow of T_1 in (2b). Second, if $y = v^*$, then $\tilde{\pi}^{-1}(y)$ is defined by (2c) if x_1^1 is a node and by (2d) or (2e) when x_1^1 is an arrow.

For the first claim of (i), note that the image of $\tilde{\pi}^{-1}$ is disjoint from E', so that $\tilde{\pi}(\tilde{\pi}^{-1}(y))$ is well-defined. Cases (1a) and (2a) define mutual inverses, as do (1b) and (2b). Case (1c) inverts (2c), (2d), (2e), and (2f) by definition (though not necessarily the other way around).

Now suppose that $y \in X_2$ is valid. If x_1^1 is a node, then $v^* = \pi(x_1^1)$ is an interior node and thus not valid. Nor is x_1^1 in the image of π^{-1}. So $\tilde{\pi}^{-1}(y)$ is distinct from x_1^1.

To show that $\tilde{\pi}^{-1}(y)$ is valid, we consider each case of the definition of $\tilde{\pi}^{-1}$ in turn.

Case (2a): y is a boundary node with $(y, \pi(x_1^1)) \in E_2$. Then $\tilde{\pi}^{-1}(y)$ is the arrow $(\pi^{-1}(y), x_1^1)$ which runs between two boundary nodes by Facts 1 and 2 and is thus valid.

Case (2b): This case is similar to Case (2a).

Case (2c): This case does not occur when y is valid by Fact 2.

Case (2d): $y = v^*$ and x_1^1 is an unstable arrow, with $\tilde{\pi}^{-1}(y) = t(x_1^1)$. Since x_1^1 is a valid boundary arrow, $t(x_1^1)$ is a boundary node. The edge $\tilde{\rho}_2(y)$ results from applying Move II to the subpath $\tilde{\rho}_1(s(x_1^1))\tilde{\rho}_1(t(x_1^1))$. Furthermore, by Fact 2, the interior neighborhoods of y and $t(x_1^1)$ are identified. Since y is valid and the break $\tilde{\rho}_1(x_1^1)$ is unstable, we conclude that $t(x_1^1)$ is valid (while $s(x_1^1)$ is not).

Case (2e): This case is similar to Case 4 when x_1^1 is stable. Note that if $y = v^*$ is valid, then x_1^1 cannot be an interior arrow.

Case (2f): If y is a valid arrow then its ends are both interior or both boundary, and it suffices to show that the same is true of $\tilde{\pi}^{-1}(y) = \pi^{-1}(y)$. This follows from Fact 1 if y is not bounded by the interior node v^*. On the other hand, if y starts or ends at v^*, then both ends of y are interior nodes, and thus y is an interior arrow. Hence $\pi^{-1}(y)$ is also an interior arrow by Fact 1, and we conclude that its ends are both interior as well. If y is a valid node and does not fall into (2a) through (2e), then $\mathrm{Star}^o(y)$ does not include v^* and thus $\mathrm{Star}^o(\pi^{-1}(y))$ is legal as well by Fact 1. Therefore $\pi^{-1}(y)$ is also a valid node. This completes (i).

For (ii), we suppose $x \in X_1$ is distinct from x_1^1 and valid. If x_1^1 is an arrow, then $E' = \{x_1^1\}$ so $x \notin E'$. If x_1^1 is a node, then E' consists of arrows between the boundary node x_1^1 and interior nodes. These arrows are not valid, so $x \notin E'$. Therefore $\tilde{\pi}(x)$ is defined.

Finally, we show that $\tilde{\pi}^{-1}(\tilde{\pi}(x)) = x$. As in (i), cases (1a) and (2a) are mutual inverses, as are (1b) and (2b). So from now on we assume x falls into (1c), with $\tilde{\pi}(x) = \pi(x)$. If $\pi(x) = \pi(x_1^1)$ as in (2c), then $x = x_1^1$ since the nodes adjacent to x_1^1 are not valid. If $\pi(x)$ falls into (2d) or (2e), then x_1^1 is a boundary arrow bounded by the valid node x and $\pi(x) = \pi(s(x_1^1)) = \pi(t(x_1^1))$. If x_1^1 is unstable (resp., stable) then we must have $x = t(x_1^1)$ (resp., $x = s(x_1^1)$) in order for $\mathrm{Star}^o(x)$ to be legal. This is consistent with the definition of $\tilde{\pi}^{-1}$ in (2d) and (2e). Finally, if $\pi(x)$ falls into (2f), then $\tilde{\pi}^{-1}(\tilde{\pi}(x)) = \pi^{-1}(\pi(x)) = x$ by Fact 1. $\square$

Paralleling our strategy for higraphs, we now populate a lower triangular matrix whose (i,j)-entry x_j^i is a valid marker on $\tilde{\rho}_j$, with $\tilde{\rho}_j = \tilde{q}(\tilde{\rho}_{j-1}, x_{j-1}^{j-1})$.

Let $(x_1^1, x_1^2, \ldots x_1^k)$ be a k-tuple of markers on a path $\tilde{\rho}_1$. For each j from 2 to k, we set

$$\tilde{\rho}_j = \tilde{q}(\tilde{\rho}_{j-1}, x_{j-1}^{j-1})$$

and

$$x_j^i = \tilde{\pi}(x_{j-1}^i)$$

for $j \leq i \leq k$. This inductive definition may break down if some x_j^i is not a valid marker.

Definition 4.13. A k-tuple $(x_1^1, x_1^2, \ldots, x_1^k)$ of valid markers on $\tilde{\rho}_1$ is *completely valid* if x_j^i is defined and valid for all $1 \leq j \leq i \leq k$.

Thus, a completely valid k-tuple $(x_1^1, x_1^2, \ldots, x_1^k)$ on $\tilde{\rho}_1$ determines a k-step gluing sequence $(x_1^1, x_2^2, \ldots, x_k^k)$ on $\tilde{\rho}_1$.

Let $(x_1^1, x_2^2, \ldots x_k^k)$ be a k-step gluing sequence on $\tilde{\rho}_1$. For each j from k down to 2, we have

$$\tilde{\rho}_j = \tilde{q}(\tilde{\rho}_{j-1}, x_{j-1}^{j-1})$$

and can thus set

$$x_{j-1}^i = \tilde{\pi}^{-1}(x_j^i)$$

for $j \leq i \leq k$. This determines a completely valid k-tuple $(x_1^1, x_1^2, \ldots x_1^k)$ of markers on $\tilde{\rho}_1$ by Lemma 4.6, which also implies that the above maps are inverses. We conclude:

Proposition 4.7. *Let $\tilde{\rho}_1$ be a path in a bigraph. The map*

$$(x_1^1, \ldots, x_1^k) \mapsto (x_1^1, \ldots, x_k^k)$$

gives a bijection between completely valid k-tuples of markers on $\tilde{\rho}_1$ and k-step gluing sequences on $\tilde{\rho}_1$.

In a bigraph, permutations of a completely valid k-tuple need not be completely valid. Indeed, in Figure 11 each set of 3 valid makers on $\tilde{\rho}_1$ is completely valid for only two out of six possible orderings, and there are precisely eight 3-step gluing sequences. However, Lemma 2.1 does extend from higraphs to bigraphs when $k = 2$.

Proposition 4.8. *Let (x_1^1, x_1^2) be a completely valid 2-tuple for a path in a bigraph.*
(i) The transposed 2-tuple (x_1^2, x_1^1) is completely valid.
(ii) The 2-step gluing sequences for (x_1^1, x_1^2) and (x_1^2, x_1^1) terminate in the same path.
(iii) For $k \geq 2$, the number of k-step gluing sequences between any two fixed paths is even.

Proof. For (i), Let (x_1^1, x_1^2) be a valid 2-tuple of markers on $\tilde{\rho}_1$ and let (y_1^1, y_1^2) be its transposition, so $y_1^1 = x_2^1$, $y_2^1 = x_1^1$. By definition, we have $x_2^2 = \tilde{\pi}_{x_1^1}(x_1^2)$ and $y_2^2 = \tilde{\pi}_{y_1^1}(y_1^2)$, where the subscript on $\tilde{\pi}$ specifies the marker at which we glued. We claim that the marker x_2^2 on $\tilde{q}(\tilde{\rho}_1, x_1^1)$ is valid if and only if the marker y_2^2 on $\tilde{q}(\tilde{\rho}_1, y_1^1)$ is valid.

By Fact 1, the claim is clear if x_1^1 and x_1^2 do not interact; that is, if the subtrees of T_1 that are contracted by gluing at x_1^1 and x_1^2 are disjoint. There are five ways in which two valid markers can interact. These are pictured in columns (a) through (e) of Figure 12. One can check that x_2^2 and y_2^2 are always both valid in all cases but (e). In column (e), x_1^1 is a boundary node, x_1^2 is a boundary arrow, and $t(x_1^2) = x_1^1$ or $s(x_1^2) = x_1^1$. In this case, Figures 13 and 14 show that x_2^2 is valid if and only if y_2^2 is valid.

For the proof of (ii), let T_3^x and T_3^y be the trees underlying the paths $\tilde{q}(\tilde{q}(\tilde{\rho}_1, x_1^1), x_2^2))$ and $\tilde{q}(\tilde{q}(\tilde{\rho}_1, y_1^1), y_2^2))$. In cases (a) through (e) of Figure 12, one can check that the trees result from contracting the same subtree T' of the tree T_1, and we therefore have an isomorphism $f : T_3^x \to T_3^y$ which identifies $\tilde{\pi}_{x_2^2}(\tilde{\pi}_{x_1^1}(v))$ with $\tilde{\pi}_{y_2^2}(\tilde{\pi}_{y_1^1}(v))$ for each $v \in V_1$. The most interesting case (e) is verified in Figure 13. Furthermore, in each case, the map f is an isomorphism of paths, since for both paths the edge over the node v^* to which T' contracts has the same source and target as the subpath of ρ_1 supported by T'. Finally, if x_1^1 and x_2^2 do not interact, then the isomorphism of paths is clear.

For the proof of (iii), note that we have shown that the set of 2-step gluing sequences between any two fixed paths admits a fixed-point free involution, given by transposition on the corresponding 2-tuple of valid markers. For $k > 2$, the set of k-step gluing sequences admits a fixed-point free involution as well, since each k-step gluing sequence factors uniquely as a $(k-2)$-step gluing sequence followed by a 2-step gluing sequence. We conclude that these sets contain an even number of elements. □

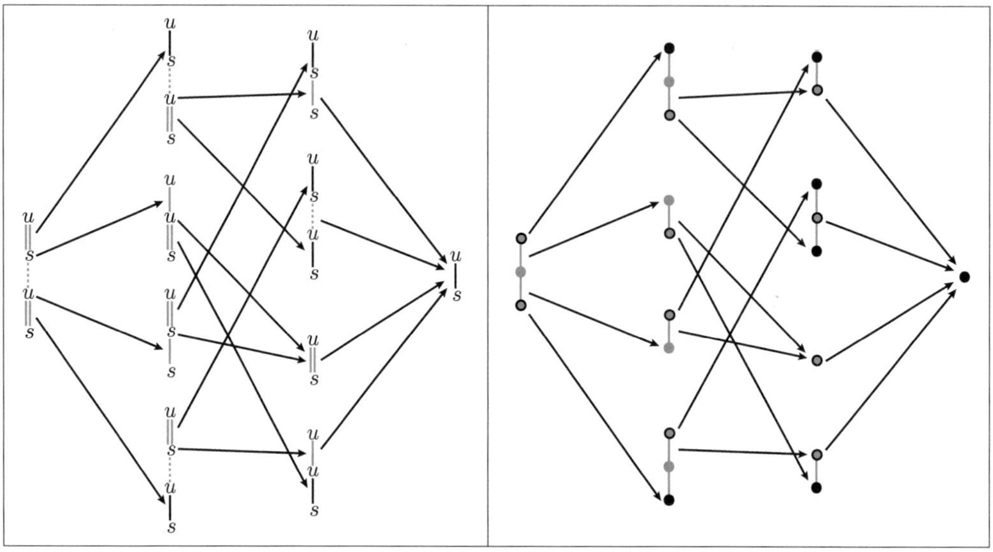

FIGURE 11. The legal path $\tilde{\rho}_1$ at left has weight 4, so all 3-step gluing sequences terminate in the path $\tilde{\rho}_4$ consisting of a single interior edge. These eight sequences are shown above, with each arrow positioned to pick out the corresponding gluing move. The four valid markers of $\tilde{\rho}_1$ are numbered on the underlying tree T_1, with 1 and 4 marking Move Ib and 2 and 3 marking Move II. Note that the 2-tuples $(1,2)$, $(2,1)$, $(3,4)$ and $(4,3)$ fall into the undefined case of column (e) of Figure 12, as illustrated in Figure 14. The eight completely valid 3-tuples on $\tilde{\rho}_1$ are $(1,3,4), (1,4,3), (2,3,4), (2,4,3), (3,1,2), (3,2,1), (4,1,2)$, and $(4,2,1)$.

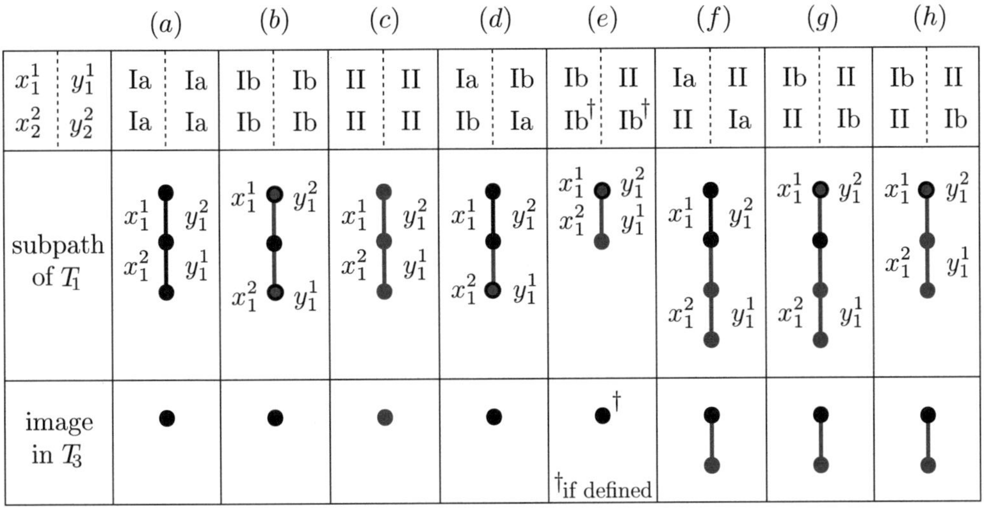

FIGURE 12. To a 2-tuple (x_1^1, x_1^2) of valid markers, we may associate the minimal undirected subpath of T_1 which includes both markers. Several such subpaths are depicted in the second row above. There are exactly five (undirected) subpaths that may arise when the contracting trees of x_1^1 and x_1^2 overlap. These are depicted in columns (a) through (e). In the first row, we depict the gluing type of the markers x_1^1, x_2^2, y_1^1, and y_2^2, where (y_1^1, y_1^2) is the transposition of (x_1^1, x_1^2). In all columns except (e), the markers x_2^2 and y_2^2 are always valid. In column (e), x_2^2 is valid if and only if y_2^2 is valid, as shown in Figures 13 and 14. If x_2^2 and y_2^2 are valid, then $\tilde{\rho}_3$ is defined and independent of gluing order, as may be easily verified in each case above. In the third row, we show the image in T_3 of the above subpath of T_1.

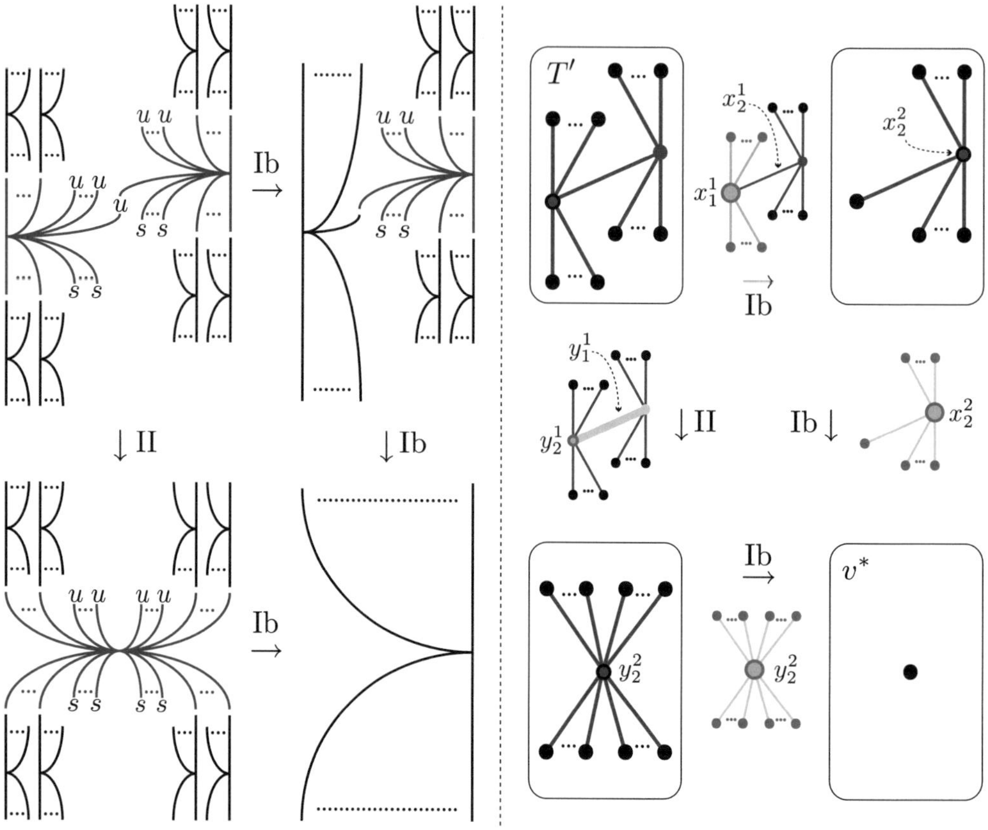

FIGURE 13. In Figures 13 and 14, we examine the case of column (e) of Figure 12 in detail. The above diagram illustrates the case when both gluing sequences (x_1^1, x_2^2) and (y_1^1, y_2^2) are defined. The subtree contracted at each step in shown in gray at right. Overall, the two sequences contract the same subtree T' of T_1 to a single node v^*, and therefore terminate in the same path $\tilde{\rho}_3$. The diagram further illustrates that x_2^2 is valid if and only if y_2^2 is valid (compare with Figure 14).

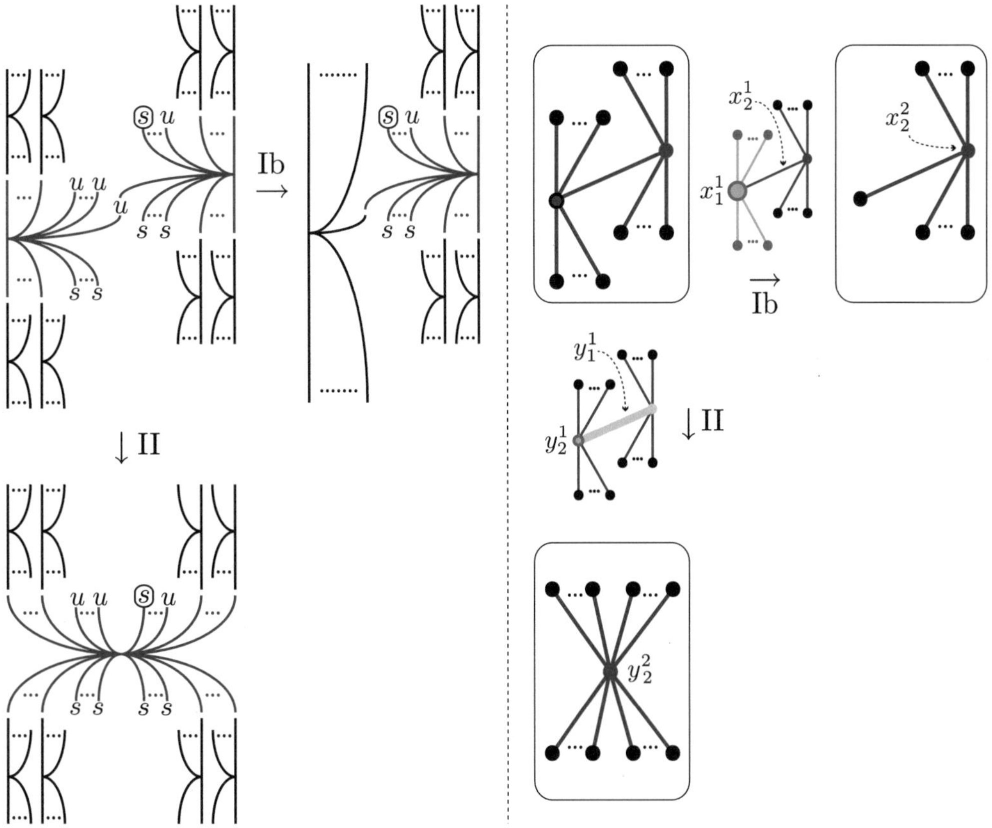

FIGURE 14. In the above diagram, both x_2^2 and y_2^2 are invalid, so that neither 2-step gluing sequence is defined. The difference between Figures 13 and 14 amounts to switching the single (encircled) vertex from unstable to stable.

Appendix: Morse homology with boundary

A. Morse homology on a closed manifold

We first recall the construction of Morse homology on a closed, smooth manifold M. Fix a Riemannian metric and Morse function $f : M \to \mathbb{R}$. The gradient flow $\phi_t : M \to M$ with respect to $-\nabla f$ determines an unstable manifold

$$U_a = \{x \in M : \lim_{t \to -\infty} \phi_t(x) = a\}$$

and stable manifold

$$S_a = \{x \in M : \lim_{t \to \infty} \phi_t(x) = a\}$$

for each critical point a. The Morse index λ_a of a is the dimension of U_a. We insist that f satisfies the *Morse-Smale* condition:

- U_a and S_b intersect transversely in M for all critical points a and b.

The manifold $U_a \cap S_b$ has dimension $\lambda_a - \lambda_b$ and is canonically identified with the moduli space of gradient trajectories from a to b:

$$\mathcal{M}(a,b) = \left\{ \gamma : \mathbb{R} \to M : \frac{d\gamma}{dt} = -\nabla f \circ \gamma, \lim_{t \to -\infty} \gamma(t) = a, \lim_{t \to +\infty} \gamma(t) = b \right\}.$$

The space $\breve{\mathcal{M}}(a,b)$ of unparameterized gradient trajectories is the quotient of $\mathcal{M}(a,b)$ by the $\mathbb{R}$-translation $\gamma(\cdot) \mapsto \gamma(\cdot + s)$, so

$$\dim \breve{\mathcal{M}}(a,b) = \lambda_a - \lambda_b - 1. \tag{17}$$

The Morse complex $C(M)$ is the $\mathbb{F}$-vector space with basis the set of critical points:

$$C(M) = \bigoplus_{a \in \mathrm{crit}(f)} \mathbb{F}a$$

The differential ∂ counts unparameterized gradient trajectories:

$$\langle \partial a, b \rangle = |\breve{\mathcal{M}}(a,b)|.$$

This notation signifies that the coefficient of b in ∂a is the number of unparameterized gradient trajectories from a to b, with the convention $|\cdot| = 0$ on infinite sets. The Morse complex is graded by Morse index and the differential has degree -1 by (17). The Morse homology $H_*(C(M), \partial)$ is isomorphic to the singular homology of M.

B. Morse homology on a manifold with boundary

Now let M be a smooth manifold with boundary. The manifold M may be identified with the quotient of a smooth double

$$\tilde{M} = M \cup_{\partial M} -M$$

by the natural involution $i : \tilde{M} \to \tilde{M}$. An i-invariant metric and Morse function on $\tilde{M}$ determine a metric and Morse function on M by restriction. For such a metric and Morse function on M, the gradient field ∇f is tangent to ∂M along the boundary. Furthermore, at each critical point on the boundary, the normal vector is an eigenvector of the Hessian $\nabla^2 f$. A boundary critical point is called *stable* or *unstable* according to whether the corresponding eigenvalue is positive or negative, respectively. Hence we may sort the critical points of f into three types: interior, (boundary-)stable, and (boundary-)unstable. We denote these types by the letters o, s, and u. If a trajectory $\gamma \in \breve{\mathcal{M}}^\partial(a,b)$ intersects the interior of M, then a must be interior or unstable and b must be interior or stable. On the other hand, if a is stable and b is unstable, then both U_a and S_b are contained in ∂M and the most we can ask is that U_a and S_b intersect transversely as submanifolds of ∂M. Indeed, we insist that f is *regular*, which means

- U_a and S_b intersect transversely in ∂M if a is stable and b is unstable.
- U_a and S_b intersect transversely in M if a is not stable or b is not unstable.

Note that regular is equivalent to Morse-Smale if $\partial M = \emptyset$.

Next we define eight linear maps between the vector spaces $C^o(M)$, $C^s(M)$, and $C^u(M)$ generated by the interior, stable, and unstable critical points, respectively:

$$\begin{aligned} \partial^o_o &: C^o(M) \to C^o(M) & \bar\partial^s_s &: C^s(M) \to C^s(M) \\ \partial^o_s &: C^o(M) \to C^s(M) & \bar\partial^s_u &: C^s(M) \to C^u(M) \\ \partial^u_o &: C^u(M) \to C^o(M) & \bar\partial^u_u &: C^u(M) \to C^u(M) \\ \partial^u_s &: C^u(M) \to C^s(M) & \bar\partial^u_s &: C^u(M) \to C^s(M) \end{aligned}$$

The maps ∂^*_* count unparameterized gradient trajectories as before, so

$$\langle \partial^*_* a, b \rangle = |\breve{\mathcal{M}}(a,b)|.$$

The maps $\bar\partial^*_*$ only count those trajectories which remain in the boundary. More precisely,

$$\langle \bar\partial^*_* a, b \rangle = |\breve{\mathcal{M}}^\partial(a,b)|$$

where $\breve{\mathcal{M}}^\partial(a,b) = \breve{\mathcal{M}}(a,b) \cap (\{\gamma : \mathbb{R} \to \partial M\}/\mathbb{R})$.

The boundary, absolute, and relative Morse complexes of M of are given by:

$$C(\partial M) = C^s(M) \oplus C^u(M) \quad C(M) = C^o(M) \oplus C^s(M) \quad C(M,\partial M) = C^o(M) \oplus C^u(M)$$

$$\partial_{\partial M} = \begin{bmatrix} \bar\partial^s_s & \bar\partial^u_s \\ \bar\partial^s_u & \bar\partial^u_u \end{bmatrix} \qquad \partial = \begin{bmatrix} \partial^o_o & \partial^u_o \bar\partial^s_u \\ \partial^o_s & \bar\partial^s_s + \partial^u_s \bar\partial^s_u \end{bmatrix} \quad \partial_{M,\partial M} = \begin{bmatrix} \partial^o_o & \partial^u_o \\ \bar\partial^s_u \partial^o_s & \bar\partial^u_u + \bar\partial^s_u \partial^u_s \end{bmatrix}$$

For each map ∂^*_*, there is a map $\tilde\partial\delta^*_*$ which counts points in the boundary of the compactification of $\breve{\mathcal{M}}(a,b)$. For each map $\bar\partial^*_*$, there is a map $\tilde\partial\delta^*_*$ which counts points in the boundary of the compactification of $\breve{\mathcal{M}}^\partial(a,b)$. These maps should be compared with the definition of δ in Figure 3.

$$\begin{aligned} \tilde\delta\partial^o_o &= \partial^o_o\partial^o_o + \partial^u_o\bar\partial^s_u\partial^o_s & \tilde\delta\bar\partial^s_s &= \bar\partial^s_s\bar\partial^s_s + \bar\partial^u_s\bar\partial^s_u \\ \tilde\delta\partial^o_s &= \partial^o_s\partial^o_o + \bar\partial^s_s\partial^o_s + \partial^u_s\bar\partial^s_u\partial^o_s & \tilde\delta\bar\partial^s_u &= \bar\partial^s_u\bar\partial^s_s + \bar\partial^u_u\bar\partial^s_u \\ \tilde\delta\partial^u_o &= \partial^o_o\partial^u_o + \partial^u_o\bar\partial^u_u + \partial^u_o\bar\partial^s_u\partial^u_s & \tilde\delta\bar\partial^u_s &= \bar\partial^s_s\bar\partial^u_s + \bar\partial^u_s\bar\partial^u_u \\ \tilde\delta\partial^u_s &= \bar\partial^u_s + \partial^o_s\partial^u_o + \bar\partial^s_s\partial^u_s + \partial^u_s\bar\partial^u_u + \partial^u_s\bar\partial^s_u\partial^u_s & \tilde\delta\bar\partial^u_u &= \bar\partial^s_u\bar\partial^u_s + \bar\partial^u_u\bar\partial^u_u \end{aligned}$$

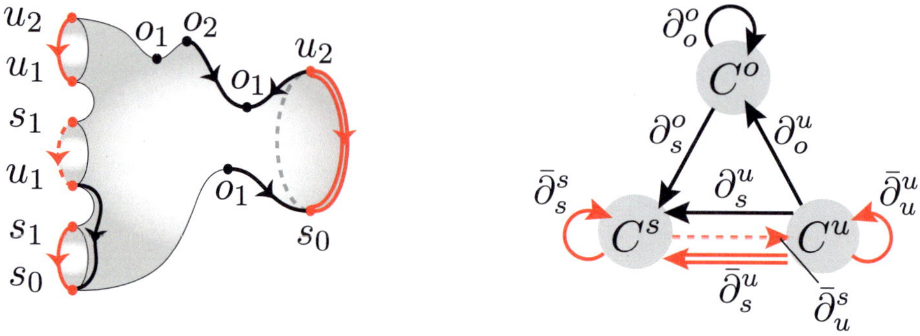

FIGURE 15. Consider the surface at left embedded in $\mathbb{R}^3$ with the induced metric and height function. Each critical point is labeled by its type and Morse index. The directed graph at right depicts the vector spaces generated by the interior, stable, and unstable critical points together with eight linear maps between them. On the surface we have highlighted a single gradient trajectory counted by each map: four interior trajectories in black and four boundary trajectories in red. The dashed red line from s to u signifies that $\bar{\partial}_u^s$ is index-preserving. The doubled red line from u to s signifies that the $\bar{\partial}_s^u$ lowers index by 2. The other six maps lower index by 1.

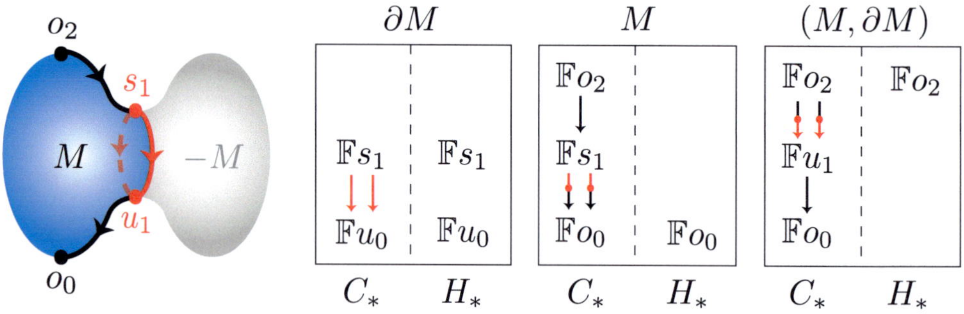

FIGURE 16. At left, the double of the 2-disk M is symmetrically embedded in $\mathbb{R}^3$. The induced metric and height function on M determine boundary, absolute, and relative Morse complexes over $\mathbb{F}$ (so two arrows are equivalent to zero arrows). In each case, the Morse homology agrees with the singular homology of the 2-disk as expected.

Alternatively, we may view the right-hand sides as relations, since each counts points in the boundary of a 1-dimensional space (itself a union of products of trajectory spaces). As in Figure 3, the structure equation takes the form

$$\begin{bmatrix} \tilde{\delta}\partial^o_o & (\tilde{\delta}\partial^u_o)\bar{\partial}^s_u + \partial^u_o(\tilde{\delta}\bar{\partial}^s_u) \\ \tilde{\delta}\partial^o_s & \delta\bar{\partial}^s_s + (\tilde{\delta}\partial^u_s)\bar{\partial}^s_u + \partial^u_s(\tilde{\delta}\bar{\partial}^s_u) \end{bmatrix} = \partial^2.$$

Since every term on the left vanishes, ∂ is indeed a differential. A similar argument applies to $\partial_{\partial M}$ and $\partial_{M,\partial M}$.

C. Relation to cell structures and CW-homology

On a closed manifold, a Morse-Smale function whose critical values increase with index induces a cell structure with one k-cell for each index-k critical point. The k-cell is given by the unstable manifold U_o of the critical point o_k. The Morse complex then coincides with the chain complex $C^{\text{cell}}(M)$ underlying CW-homology.

A similar relationship should hold for a manifold with boundary, motivated by the local picture of passing a critical level in Figure 17. We suggest that a regular function whose critical values increase with index induces a cell structure with:

- a k-cell $U_o \subset M$ for each interior critical point o_k.
- a k-cell $U_s \subset \partial M$ for each stable critical point s_k.
- a k-cell $U_u \subset M$ and a $(k-1)$-cell $\partial U_u \subset \partial M$ for each unstable critical point u_k.

This should follow from the argument in the closed case adapted to the double $\tilde{M}$, though we do not attempt this here.

Assuming the above, for each unstable critical point, the $(k-1)$-cell ∂U_u occurs with multiplicity 1 in the boundary of the k-cell U_u. There is therefore a single arrow between each pair of unstable generators in $C^{\text{cell}}_*(M)$ and a "cancellation" homotopy equivalence to a smaller complex $C'_*(M)$ generated by the interior and stable cells. The components of the differential on $C'_*(M)$ running from stable to interior generators are induced by "zig-zags" in $C^{\text{cell}}_*(M)$ involving the algebraically-cancelled pairs of unstable generators. A short exercise reveals that the complex $C'_*(M)$ is naturally identified with the Morse complex. In Figure 18, we illustrate this relationship for the example in Figure 16. A similar exercise applies to the relative and boundary complexes without cancellation. In summary:

$$\begin{aligned} C^{\text{cell}}_*(M) \xrightarrow{\sim} C'_*(M) &= C^{\text{morse}}_*(M) \\ C^{\text{cell}}_*(M, \partial M) &= C^{\text{morse}}_*(M, \partial M) \\ C^{\text{cell}}_*(\partial M) &= C^{\text{morse}}_*(\partial M) \end{aligned}$$

In this way, the rather strange Morse differentials arise naturally from the cellular perspective.

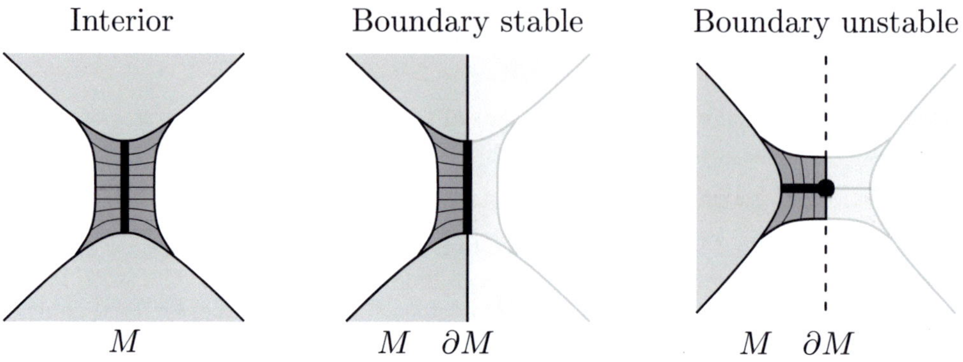

FIGURE 17. The local level-set of M just about a critical value for each type of critical point, with the mirror of M faded.

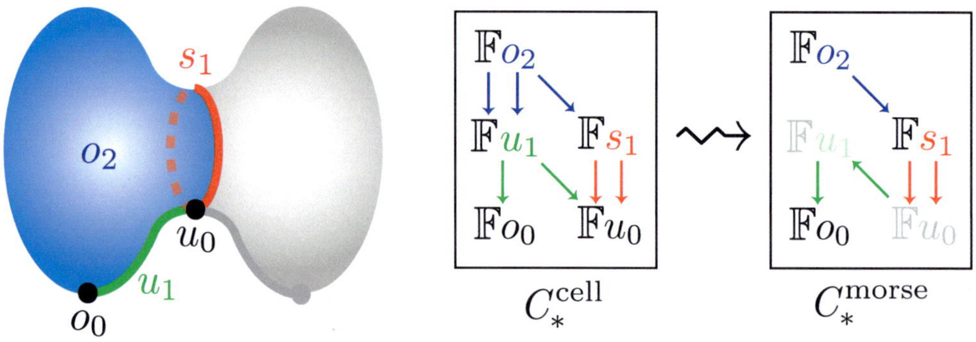

FIGURE 18. We consider the example in Figure 16 from the cellular perspective. The cellular chain complex is generated by five cells and the differential encodes the algebraic multiplicities of the attaching maps. The Morse complex results from algebraically canceling the pair of unstable cells.

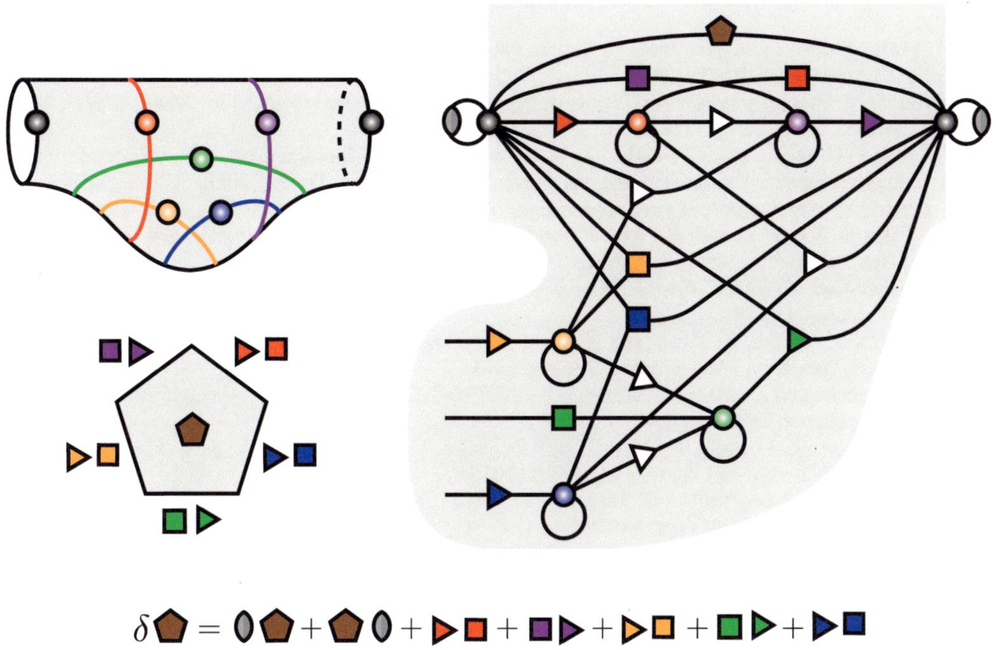

FIGURE 19. The proof of the surgery exact triangle for $\widecheck{HM}$ involves a map $\check{G}$ defined by counting monopoles on a cobordism over a family of metrics parameterized by a pentagon. The edges of the pentagon correspond to five hypersurfaces which intersect as illustrated at top left (compare with Figure 2 in [10]). The map $\check{G}$ corresponds to the edge labelled by the brown pentagon in the higraph, which has 7 vertices and and 23 edges. In [10], the four aligned vertices are type 2 ($\widecheck{HM}$) and the remaining three vertices are type 1 ($\widehat{HM}$). The path DGA on the bigraph (which need not be shown!) determines the terms of all monopole maps represented by edges in the higraph, and the structure equation implies that that $\check{G}$ satisfies the relation at the bottom of the figure (this agrees with Proposition 5.5 in [10] because the yellow and blue triangles are zero maps for geometric reasons). Note that since this higraph is tree-like, the assignment $\psi \equiv 1$ is balanced as well, leading to the surgery triangle for $\widehat{HM}$. The key point here is that, due to Theorem 4.3, one can justifiably work on the level of higraphs, as though all monopoles in life were irreducible after all.

References

[1] M. Abouzaid, A topological model for the Fukaya category of plumbings, *J. Diff. Geom.*, 87(1):1–80, 2011.

[2] M. Akaho, Floer's chain complexes for Lagrangian submanifolds in symplectic manifolds with concave ends, 2003, arXiv:math/0310166v1.

[3] J. Bloom, A link surgery spectral sequence in monopole Floer homology, *Adv. Math.*, 226(4):3216–3281, 2011.

[4] V. Colin, P. Ghiggini, and K. Honda, Equivalence of Heegaard Floer homology and embedded contact homology via open book decompositions, *Proc. Nat. Acad. Sci.*, 108(20), 2011.

[5] S. Devadoss, A realization of graph-associahedra, *Discrete Math.*, 309:271–276, 2009.

[6] K. Fukaya, Floer homology for 3 manifold with boundary I, *Unpublished notes*, 1997, http://www.math.kyoto-u.ac.jp/~fukaya/bdrt1.pdf.

[7] K. Fukaya and Y.-G. Oh, Zero-loop open strings in the cotangent bundle and Morse homotopy, *Asian J. Math*, 1(1), March 1997.

[8] K. Fukaya, Y.-G. Oh, H. Ohta, and K. Ono, *Lagrangian intersection Floer theory - anomaly and obstruction*, American Mathematical Society, 2010.

[9] P. Kronheimer and T. Mrowka, *Monopoles and Three-Manifolds*, New Mathematical Monographs. Cambridge University Press, 2007.

[10] P. Kronheimer, T. Mrowka, P. Ozsváth, and Z. Szabó, Monopoles and lens space surgeries, *Ann. of Math. (2)*, 165(2):457–546, 2007, math.GT/0310164.

[11] C. Kutluhan, Y.-J. Lee, and C. H. Taubes, HF=HM I : Heegaard Floer homology and SeibergWitten Floer homology, 2010, arXiv:1007.1979.

[12] Y. Lekili and T. Perutz, Fukaya categories of the torus and Dehn surgery, *Proc. Nat. Acad. Sci.*, 108(20), 2011, arXiv:1102.3160.

[13] Y. Lekili and T. Perutz, Arithmetic mirror symmetry for the 2-torus, 2012, arXiv:1211.4632.

[14] R. Lipshitz, P. S. Ozsváth, and D. P. Thurston, Tour of bordered Floer theory, *Proc. Nat. Acad. Sci.*, 108(20), 2011.

[15] P. Seidel, *Fukaya categories and Picard-Lefschetz theory*, Zurich Lectures in Advanced Mathematics. European Mathematical Society, 2008.

Department of Mathematics, Massachusetts Institute of Technology, Cambridge, MA 02139
E-mail address: jbloom@math.mit.edu

Proceedings of 19th Gökova
Geometry-Topology Conference
pp. 89 – 110

Instantons in G_2 manifolds from J-holomorphic curves in coassociative submanifolds

Naichung Conan Leung, Xiaowei Wang and Ke Zhu

ABSTRACT. In G_2 manifolds, 3-dimensional associative submanifolds (instantons) play a role similar to J-holomorphic curves in symplectic geometry. In [21], instantons in G_2 manifolds were constructed from regular J-holomorphic curves in coassociative submanifolds. In this exposition paper, after reviewing the background of G_2 geometry, we explain the main ingredients in the proofs in [21]. We also construct new examples of instantons.

1. Introduction

A G_2-manifold M is a Riemannian manifold of dimension seven equipped with a non-trivial parallel 2-fold *vector cross product* (VCP) $\times$. If the VCP is 1-fold instead, namely a Hermitian complex structure, then the manifold is a Kähler manifold X.[1] Note that every 2-fold VCP comes from the restriction of the algebra product on the octonions $\mathbb{O}$ or quaternions $\mathbb{H}$ to its imaginary part. Thus the only submanifolds in M preserved by $\times$ are of dimension three and they are called *instantons* or *associative submanifolds*, which are analogs to holomorphic curves in Kähler manifolds.

In Physics, G_2-manifolds are internal spaces for compactification in M-theory in eleven dimensional spacetimes, similar to the role of *Calabi-Yau threefolds* X in string theory in ten dimensional spacetimes. Instantons in string theory are holomorphic curves Σ in X with the natural boundary condition being $\partial\Sigma$ lies inside a Lagrangian submanifold L in X. Very roughly speaking, the Fukaya category $Fuk(X)$ is defined by counting holomorphic disks with Lagrangian boundary condition. The homological mirror symmetry (HMS) conjecture of Kontsevich says that $Fuk(X)$ is equivalent to the derived category $D^b(X^\vee)$ of coherent sheaves of the mirror manifold $X^\vee$. The proof of this conjecture and its generalizations in many specific cases relies on the work of Fukaya-Oh [8] when $X = T^*L$ and the Lagrangian L_t is the graph of an exact one form df scaled by small $t \in \mathbb{R}$. They showed that holomorphic disks with boundary in $L \cup L_t$ one to one correspond to gradient flow lines of f, which are in fact instantons in quantum mechanics according to Witten's Morse theory [31]. (Fukaya-Oh [8] actually proved the case of k

Key words and phrases. G_2 manifolds, instantons, J-holomorphic curves, coassociative submanifolds.

[1]Throughout this paper M is a G_2-manifold and X is a symplectic manifold possibly with extra structures, like Kähler or CY.

Lagrangians $L_t^j = tdf_j$ $(j = 1, 2, \cdots k)$, where holomorphic polygons bounding on these Lagrangians correspond to gradient flow trees of the Morse functions $\{f_j\}_{j=1}^k$.)

In [21] we proved a corresponding result for instantons in any G_2-manifold with boundary in the coassociative submanifold $C \cup C_t$. Here the family C_t is constructed by a self-dual harmonic two form ω on the four manifold C. We assume that ω is non-degenerate, thus defining an almost complex structure J on C. Our main result, Theorem 13, gives a correspondence between such instantons in M and J-holomorphic curves Σ in C. Thus the number of instantons in M is related to the Seiberg-Witten invariant of C by the celebrated work of Taubes on GW=SW ([26], [28], [27]). We suspect this holds true without the need of the non-degeneracy of ω.

Our result is similar to the $k = 2$ case of Fukaya-Oh [8], however the analysis involved in the proof is essentially different from theirs in the following 3 aspects.

(1) The instanton equation is on 3-*dimensional* domains, and there is no analogous way of finding associative submanifold by constructing *associative maps* as was done in constructing J-holomorphic curves thanks to the *conformality* of Cauchy-Riemann equation. So we have to deform *submanifolds* rather than *maps* as in the Lagrangian Floer theory. A good choice of normal frames of submanifolds turns out to be essential.

(2) The instanton equation is *more nonlinear* than the Cauchy-Riemann equation for J-holomorphic curves. It is a first order PDE system involving cubic terms $\frac{\partial V^i}{\partial x^1}\frac{\partial V^j}{\partial x^2}\frac{\partial V^k}{\partial x^3}$, while the Cauchy-Riemann operator $\frac{\partial u}{\partial \tau} + J(u)\frac{\partial u}{\partial t}$ has no product of derivative terms. Consequently, the needed *quadratic estimate* appears to be unavailable in the $W^{1,p}$ setting, as for $V^i, V^j, V^k \in W^{1,p}$, $\frac{\partial V^i}{\partial x^1}\frac{\partial V^j}{\partial x^2}\frac{\partial V^k}{\partial x^3} \notin L^p$ in general. So instead we use the *Schauder* $(C^{1,\alpha})$ *setting*, which creates new complications (Subsection 3.2.4).

(3) The linearized instanton equation is *more weakly coupled* than the Cauchy-Riemann equation. It is a Dirac type equation for spinors $(u, v) \in \mathbb{S}^+ \oplus \mathbb{S}^-$ where u and v play the role of the real and imaginary parts in Cauchy-Riemann equations, but the interrelation between ∇u and ∇v becomes weaker. This causes several difficulties in the $C^{1,\alpha}$ estimates (See comments below (14)).

Besides above difficulties, our domains are $[0, \varepsilon] \times \Sigma$ for compact Riemannian surfaces Σ, and they *collapse* to Σ as $\varepsilon \to 0$, causing *lack of uniform ellipticity,* which in turn creates difficulty to obtain a uniform right inverse bound needed in gluing arguments. This also occurs in Proposition 6.1 of Fukaya-Oh [8] in the $W^{1,p}$ setting, but in our $C^{1,\alpha}$ setting the boundary estimates become more subtle.

To deal with these difficulties, our paper [21] becomes rather technical. Therefore in this article we give an outline of the main arguments. The organization is as follows. In Section 2, we review the background of G_2 geometry, instantons, coassociative boundary condition and give the motivations of counting instantons. In Section 3 we state the main theorem in [21] and explain the main ingredients in the proof. In Section 4 we apply the theorem to construct new examples of instantons and discuss possible generalizations.

Acknowledgement. We thank S.-T. Yau for comments on our paper [21], and Baosen Wu for useful discussions on complex surfaces. The third author thanks Clifford Taubes for his interest and support. The work of N.C. Leung was substantially supported by a grant from the Research Grants Council of the Hong Kong Special Administrative Region, China (Project No. CUHK401411). The work of K. Zhu was partially supported by the grant of Clifford Taubes from the National Science Foundation.

2. Review on G_2-geometry

2.1. G_2 manifolds

G_2 manifolds are 7-dimensional Riemannian manifolds (M, g) with a parallel *vector cross product* (VCP) $\times$, i.e., for $u, v \in T_pM$ there is a VCP $\times$ such that

(i) $u \times v$ is perpendicular to both u and v.
(ii) $|u \times v| = |u \wedge v| =$ Area of the parallelogram spanned by u and v,

and $\times$ is invariant under the parallel transport of the Levi-Civita connection ∇.

In the octonion algebra $\mathbb{O}$, we can construct a VCP $\times$ on Im $\mathbb{O}$ as follows:

$$u \times v = \operatorname{Im}(\overline{v}u), \qquad (1)$$

where $\overline{v}$ is the conjugate of v. The same formula for $\mathbb{H}$ gives another VCP on $\operatorname{Im}\mathbb{H}$, which is indeed the useful vector product on $\mathbb{R}^3$. Together, they form the complete list of VCP because the normed algebra structures can be recovered from the VCP structures. In particular,

$$Aut\left(\mathbb{R}^7, \times\right) = Aut_{alg}(\mathbb{O}).$$

Hence a G_2 manifold is simply a 7-dimensional Riemannian manifold (M, g) with holonomy group inside the exceptional Lie group $G_2 = Aut_{alg}(\mathbb{O})$.

Equivalently, G_2 manifolds are 7-dimensional Riemannian manifolds (M, g) with a *nondegenerate* 3-*form* Ω such that $\nabla\Omega = 0$. The relation between Ω and $\times$ is

$$\Omega(u, v, w) = g(u \times v, w), \quad \text{for } u, v, w \in T_pM.$$

Example 1 (Linear case). *The G_2 manifold* $\operatorname{Im}\mathbb{O} \simeq \mathbb{R}^7$*: Let*

$$\mathbb{R}^7 \simeq \operatorname{Im}\mathbb{O} \simeq \operatorname{Im}\mathbb{H} \oplus \mathbb{H} = \{(x_1\mathbf{i} + x_2\mathbf{j} + x_3\mathbf{k}, x_4 + x_5\mathbf{i} + x_6\mathbf{j} + x_7\mathbf{k})\},$$

and the standard basis $e_i = \frac{\partial}{\partial x_i}$ $(i = 1, 2, \cdots, 7)$*. The vector cross product $\times$ for u, v in* $\operatorname{Im}\mathbb{O}$ *is defined by* (1)*. The standard G_2 3-form $\Omega = \Omega_0$ is*

$$\Omega_0 = \omega^{123} - \omega^{167} - \omega^{527} - \omega^{563} - \omega^{154} - \omega^{264} - \omega^{374},$$

where $\omega^{ijk} = dx^i \wedge dx^j \wedge dx^k$.

Example 2 (Product case). *$M = X \times S^1$ is a G_2 manifold if and only if X is a Calabi-Yau threefold. The G_2 3-form Ω of X is related to the holomorphic volume form Ω_X and the Kähler form ω_X of X as follow*

$$\Omega = \operatorname{Re}\Omega_X + \omega_X \wedge d\theta,$$

where $d\theta$ is the standard angular-form on S^1.

So far all compact irreducible G_2 manifolds are constructed by solving the nonlinear PDE of the G_2 metric using the implicit function theorem, including (i) resolving orbifold singularities by Joyce [17], [18] and (ii) twisted connected sum by Kovalev [19] and Corti-Haskins-Nordstrom-Pacini [7].

We remark a useful construction of local G_2 frames.

Remark 3 (Cayley-Dickson construction). *A convenient basis in $\operatorname{Im}\mathbb{O}$ can be constructed inductively (Lemma A.15 in [13]): Given orthogonal unit vectors v_1 and v_2, we define $v_3 := v_1 \times v_2$. Then we take any unit vector $v_4 \perp span\{v_1, v_2, v_3\}$, and let $v_{i+4} := v_i \times v_4$ for $i = 1, 2, 3$. Then $\{v_i\}_{i=1}^{7}$ is an orthonormal basis of $\operatorname{Im}\mathbb{O}$ satisfying the same vector cross product relation as the standard basis $\{e_i\}_{i=1}^{7}$ in $\operatorname{Im}\mathbb{O}$.*

This basis $\{v_i\}_{i=1}^{7}$ is useful in many local calculations in [21] (see also [3]), where $\{v_1, v_2\}$ span the tangent spaces of a J-holomorphic curve.

2.2. Instantons (associative submanifolds)

A 3-dimensional submanifold A in a G_2 manifold M is called an **associative submanifold** (or an **instanton**) if its tangent space TA is closed under $\times$. This notion was introduced by Harvey-Lawson ([13], see also [20]). The following are two other equivalent conditions for A to be an associative submanifold: (1) A is calibrated by Ω, namely $\Omega|_A = dv_A$; (2) $\tau|_A = 0 \in \Omega^3(A, T_M|_A)$ where $\tau \in \Omega^3(M, TM)$ is defined by the following equation

$$g(\tau(u, v, w), z) = (*\Omega)(u, w, w, z) \tag{2}$$

for any u, v, w and $z \in T_pM$, where $*\Omega \in \Omega^4(M)$ is the Hodge-$*$ of Ω. This measurement $\tau|_A$ of associativity is important for perturbing almost instantons to an instanton.

Example 4. *In $\operatorname{Im}\mathbb{O} \simeq \operatorname{Im}\mathbb{H} \oplus \mathbb{H}$, the subspace $\operatorname{Im}\mathbb{H} \oplus \{0\}$ is an instanton. Explicitly τ is ((5.4) in [24])*

$$\begin{aligned}
\tau = {} & \left(\omega^{256} - \omega^{247} + \omega^{346} - \omega^{357}\right)\partial_1 + \left(\omega^{156} - \omega^{147} - \omega^{345} + \omega^{367}\right)\partial_2 \\
& + \left(\omega^{245} - \omega^{267} - \omega^{146} - \omega^{157}\right)\partial_3 + \left(\omega^{567} - \omega^{127} + \omega^{136} - \omega^{235}\right)\partial_4 \\
& + \left(\omega^{126} - \omega^{467} + \omega^{137} + \omega^{234}\right)\partial_5 + \left(\omega^{457} - \omega^{125} - \omega^{134} + \omega^{237}\right)\partial_6 \\
& + \left(\omega^{124} - \omega^{456} - \omega^{135} - \omega^{236}\right)\partial_7,
\end{aligned} \tag{3}$$

where $\partial_i = \frac{\partial}{\partial x^i}$. For a map $V = \left(V^4, V^5, V^6, V^7\right) : \operatorname{Im}\mathbb{H} \to \mathbb{H}$, we let

$$A = graph(V) = \{(x_1, x_2, x_3, V(x_1, x_2, x_3)) \,|\, (x_1, x_2, x_3) \in \operatorname{Im}\mathbb{H}\}.$$

By (3), *the condition* $\tau|_A = 0$ *becomes* $V^*\tau = 0$ *in coordinates* (x_1, x_2, x_3), *and we see the instanton equation for* A *is a* first order cubic PDE *system involving product terms* $\frac{\partial V^i}{\partial x^1}\frac{\partial V^j}{\partial x^2}\frac{\partial V^k}{\partial x^3}$.

Example 5. *For any holomorphic curve* Σ *in a Calabi-Yau manifold* X, $\Sigma \times S^1$ *is an instanton in the* G_2 *manifold* $X \times S^1$.

The *deformations* of an instanton A in a G_2 manifold M are governed by a *twisted Dirac operator* on its normal bundle $N_{A/M}$, regarded as a Clifford module over TA with $\times$ as the Clifford multiplication.

Theorem 6 ([24]). *Infinitesimal deformations of instantons at* A *are parameterized by the space of harmonic twisted spinors on* A, *i.e., kernel of the twisted Dirac operator.*

Example 7. *For the instanton* $\operatorname{Im}\mathbb{H} \oplus \{0\} \subset \operatorname{Im}\mathbb{O}$ *in Example 4, regarding sections* $V = V^4 + \mathbf{i}V^5 + \mathbf{j}V^6 + \mathbf{k}V^7$ *in its normal bundle as* $\mathbb{H}$*-valued functions, then the twisted Dirac operator is* $\mathcal{D} = \mathbf{i}\nabla_1 + \mathbf{j}\nabla_2 + \mathbf{k}\nabla_3$, *and*

$$\begin{aligned}\mathcal{D}V = &-\left(V_1^5 + V_2^6 + V_3^7\right) + \mathbf{i}\left(V_1^4 + V_3^6 - V_2^7\right) \\ &+ \mathbf{j}\left(V_2^4 - V_3^5 + V_1^7\right) + \mathbf{k}\left(V_3^4 + V_2^5 - V_1^6\right). \end{aligned} \tag{4}$$

Remark 8. *Constructing instantons in* G_2 *manifolds in general is difficult, partly because the deformation theory can be obstructed. Our main theorem in* [21] *provides a construction of instantons from* J*-holomorphic curves in coassociative submanifolds. Other constructions are in* [7].

2.3. Boundary value problem on coassociative submanifolds

Coassociative submanifolds are analogues of Lagrangian submanifolds in symplectic geometry.

Definition 9. *A 4-dimensional submanifold* C *in a* G_2 *manifold* (M, g, Ω) *is called coassociative if* $\Omega|_C = 0$. *Equivalently,* C *is calibrated by* $*\Omega$.

Example 10. *When* $M = X \times S^1$ *is a product with* X *a Calabi-Yau threefold, then (i)* $C = L \times S^1$ *is coassociative if and only if* $L \subset X$ *is a special Lagrangian submanifold* L *of phase* $\pi/2$ *and (ii)* $C = S \times \{\theta\}$ *is coassociative if and only if* $S \subset X$ *is a complex surface.*

Deformations of a coassociative submanifold C is studied by McLean [24]: Infinitesimally, they are parameterized by $H_+^2(C)$, the space of self-dual harmonic 2-forms on C, and they are always unobstructed.

Finding instantons with boundaries on coassociative submanifolds is an elliptic problem, similar to finding J-holomorphic curves with boundaries on Lagrangian submanifolds in symplectic geometry.

Theorem 11 (Theorem 4.2, [10]). *The linearization of the instanton equation on an instanton A with boundaries Σ lying on a coassociative submanifold C is an elliptic Fredholm operator, with the Fredholm index given by*

$$\int_{\Sigma} c_1\left(N_{\Sigma/C}\right) + 1 - g\left(\Sigma\right).$$

We have the following orthogonal decomposition of TM along Σ,

$$TM|_{\Sigma} = \mathbb{R}\langle n\rangle \oplus T\Sigma \oplus N_{\Sigma/C} \oplus N_{\langle C,n\rangle/M}|_{\Sigma},$$

where n be the unit inward normal vector field of Σ in A. We call $N_{\Sigma/C}$ the *"intrinsic"* normal bundle of Σ in C and $N_{\langle C,n\rangle/M}$ the *"extrinsic"* normal bundle defined as the orthogonal complement of $TC\oplus\langle n\rangle$ in TM. In [10], Gayet-Witt showed that (1) $T\Sigma$, $N_{\Sigma/C}$ and $N_{\langle C,n\rangle/M}|_{\Sigma}$ are complex line bundles with respect to the almost complex structure $J_n := n\times$ and (2) $N_{\Sigma/C} \otimes_{\mathbb{C}} \wedge^{0,1}_{\mathbb{C}} (T^*\Sigma) \simeq N^*_{\langle C,n\rangle/M}$ as complex line bundles over Σ which is obtained by changing the tensor product $\otimes$ to the vector cross product $\times$, and used the metric to identify conjugate bundle with dual bundle. This relationship between intrinsic and extrinsic geometry of Σ in C is needed ([21]) in order to show two natural Dirac operators on Σ agree which is needed in subsection 3.2.5 for the proof of our main theorem.

The following table gives an interesting comparison between G_2-geometry and symplectic geometry.

G_2 **manifold**	**Symplectic manifold**
nondegenerate 3-form Ω, $d\Omega = 0$	nondegenerate 2-form ω, $d\omega = 0$
vector cross product $\times$	almost complex structure J
$\Omega(u,v,w) = g(u\times v, w)$	$\omega(u,v) = g(Ju, v)$
instanton A TA preserved by$\times$ A calibrated by Ω	J-holomorphic curve Σ $T\Sigma$ preserved by J Σ calibrated by ω
coassociative submanifold C $\Omega\|_C = 0$ and $\dim C = 4$	Lagrangian submanifold L $\omega\|_L = 0$ and $\dim L = \frac{1}{2}\dim X$

2.4. Intersection theory?

Intersection theory of Lagrangian submanifolds is an essential part of symplectic geometry. By counting the number of holomorphic disks bounding intersecting Lagrangian submanifolds, Floer and others defined the celebrated Floer homology theory. It plays an important role in mirror symmetry for Calabi-Yau manifolds and string theory in physics.

In M-theory, Calabi-Yau threefolds are replaced by G_2-manifolds. The analogs of holomorphic disks (resp. Lagrangian submanifolds) are instantons or associative submanifolds (resp. coassociative submanifolds or branes) in M [20]. The problem of counting instantons has been discussed by many physicists. For example Harvey and Moore [14] discussed the mirror symmetry aspects of it; Aganagic and Vafa [2] related it to the open Gromov-Witten invariants for local Calabi-Yau threefolds; Beasley and Witten [4] argued that one should count instantons using the Euler characteristic of their moduli spaces.

On the mathematical side, the compactness issues of the moduli of instantons is a very challenging problem because the bubbling-off phenomena of (3-dimensional) instantons is not well understood. This makes it very difficult to define an honest invariant by counting instantons. Note that the Fredholm theory for instantons with coassociative boundary conditions has been set up [10].

In symplectic geometry, Fukaya and Oh [8] considered two *nearby* Lagrangian submanifolds L and L_t, where L_t is the graph of a closed 1-form α on L scaled by a small t, then holomorphic disks bounding them with *small volume* is closely related to gradient flow lines of α. They actually deal with J-holomorphic polygons bounding k Lagrangians and need to smooth vertex singularities on gradient flow trees when $k \geq 3$. For simplicity we only state their result for the $k = 2$ case here.

Given a closed 1-form α on L, $L_t :=$graph$(t\alpha)$ is a Lagrangian submanifold in T^*L near the zero section L. By Weinstein's neighborhood theorem, a small tubular neighborhood of a Lagrangian L in a symplectic manifold can be identified as a tubular neighborhood of L in T^*L.

Theorem 12. ([8]) *For any compact Riemannian manifold (L, g), T^*L has a natural almost Kähler structure (ω, J_g). Let $L_t :=$graph$(t\alpha)$ be a Lagrangian submanifold in T^*L for a closed 1-form α on L. There is a constant $\varepsilon_0 > 0$, such that for any $t \in (0, \varepsilon_0]$, there is a 1-1 correspondence between J_g-holomorphic curves bounding $L \cup L_t$ and gradient flow lines of α on L.*

We want to build the following analogue: instantons A bounding $C \cup C_t$ are in 1-1 correspondence to J_n-holomorphic curves Σ on C.

Symplectic manifold M	**G_2 manifold M**
Lagrangian submanifolds L and L_t	coassociative submanifolds C and C_t
J-holo. curve bounding $L \cup L_t$	instanton A bounding $C \cup C_t$
gradient flow line of α on L	J_n-holomorphic curve Σ on C

The meaning of J_n, C_t and the precise statment of our result will be explained in the next section.

3. Main Theorem: instantons from J-holomorphic curves

3.1. Statement of the main theorem

Let $\mathcal{C} = \cup_{0 \leq t \leq \varepsilon} C_t$ be a family of coassociative manifolds C_t in a G_2-manifold M, regarded as a deformation of $C = C_0$ along the normal vector field $n := \frac{dC_t}{dt}|_{t=0}$. Then $\iota_n \Omega$

is a self-dual harmonic 2-form[2] on C by McLean's Theorem on deformations of coassociative submanifolds (Section 2.3). In particular $\omega_n := \iota_n \Omega$ defines a *symplectic structure* on $C \backslash n^{-1}(0)$ as $\omega_n \wedge \omega_n = *\omega_n \wedge \omega_n = |\omega_n|^2 \, dv_C$ is nonzero outside $\{n = 0\}$. Furthermore $J_n := \frac{n}{|n|} \times$ defines a compatible *almost complex structure* on $\left(C \backslash n^{-1}(0), \omega_n\right)$. When n has no zeros, we have the following main theorem in [21]

Theorem 13. *Suppose that (M, Ω) is a G_2-manifold and $\{C_t\}$ is an one-parameter smooth family of coassociative submanifolds in M. When $\iota_n \Omega \in \Omega^2_+(C_0)$ is nonvanishing, then*

(1) *(Proposition 6) If $\{\mathtt{A}_t\}$ is any one-parameter family of associative submanifolds (i.e. instantons) in M satisfying*

$$\partial \mathtt{A}_t \subset C_t \cup C_0, \ \lim_{t \to 0} \mathtt{A}_t \cap C_0 = \Sigma_0 \text{ in the } C^1\text{-topology},$$

then Σ_0 is a J_n-holomorphic curve in C_0.

(2) *(Theorem 24) Conversely, every regular J_n-holomorphic curve Σ_0 (namely those for which the linearization of $\overline{\partial}_{J_n}$ on Σ_0 is surjective) in C_0 is the limit of a family of associative submanifolds $\mathtt{A}_t$'s as described above.*

Our results are similar to those in Fukaya-Oh [8] and the proofs also share some similarities: relating the Fredholm regular property of higher dimensional linearized instanton equations to lower dimensional ones; necessity to deal with the lack of uniform ellipticity as the domain collapses when $\varepsilon \to 0$; using the periodic reflection technique to "thicken" the collapsing domain to achieve a uniform right inverse estimate in the $W^{1,p}$ setting.

However the proof in our case has more difficulties than those needed in the $k = 2$ case of [8], as explained in the introduction. For $k \geq 3$, [8] contains difficulties we have not encountered here: to find the local models of the singularities of degenerating J-holomorphic polygons and resolve them.

Remark 14. *Given any Riemann surface $\Sigma \subset M$, it can always be thickened to an instanton by the Cartan-Kähler theory ([13],[10]). However its boundary may not lie inside any coassociative submanifold (see [10]). In our case, we produce an instanton A_ε with boundary in the coassociative submanifold $C \cup C_\varepsilon$, but $\partial A_\varepsilon \cap C$ is only close but not equal to Σ.*

3.2. Main ingredients of the proof

3.2.1. Formulating the instanton equation near an almost instanton

We first produce an almost instanton with boundaries on $C_0 \cup C_\varepsilon$. Let $\varphi : [0, \varepsilon] \times C \to M$ be a parametrization of the family of coassociative submanifolds $\{C_t\}_{0 \leq t \leq \varepsilon}$. Under the assumptions that $n = \frac{dC_t}{dt}|_{t=0}$ is nonvanishing and Σ is a J_n-holomorphic curve in $C = C_0$, we define

$$A_\varepsilon := [0, \varepsilon] \times \Sigma, \text{ and } A'_\varepsilon = \varphi(A_\varepsilon) \subset M,$$

[2]Here ι_n is the contraction of a form by the vector field n.

then A'_ε is an *almost instanton* with $\partial A'_\varepsilon \subset C_0 \cup C_\varepsilon$ in the sense that $\left|\tau|_{A'_\varepsilon}\right|_{C^0}$ is small. Recall that a 3-dimensional submanifold $A \subset M$ is an instanton if and only if $\tau|_A = 0$, where $\tau \in \Omega^3(M, TM)$ is defined in (2). The reason for the smallness is that, at the point $p \in \varphi(\{0\} \times \Sigma) \subset A'_\varepsilon$, $T_p A'_\varepsilon$ is associative by the J_n-holomorphic property of Σ so $\tau|_{T_p A'_\varepsilon} = 0$, and any point q on A'_ε has ε-order distance to $\varphi(\{0\} \times \Sigma)$ while $\tau|_{T_q A'_\varepsilon}$ smoothly depends on q.

Next we formulate $\tau|_A$ as a nonlinear map F_ε on the space $\Gamma\left(A'_\varepsilon, N_{A'_\varepsilon/M}\right)$ of sections of the *normal bundle* $N_{A'_\varepsilon/M}$ *of* A'_ε, in particular $F_\varepsilon(V) = 0$ if and only if

$$\mathtt{A}_\varepsilon(V) := (\exp V)(\mathtt{A}'_\varepsilon) \tag{5}$$

is an instanton, where $\exp V : \mathtt{A}'_\varepsilon \to M$.

To do this, we let

$$C^\alpha\left(\mathtt{A}'_\varepsilon, N_{\mathtt{A}'_\varepsilon/M}\right) = \left\{V \in \Gamma\left(\mathtt{A}'_\varepsilon, N_{\mathtt{A}'_\varepsilon/M}\right) | V \in C^\alpha\right\},$$
$$C^{1,\alpha}_-\left(\mathtt{A}'_\varepsilon, N_{\mathtt{A}'_\varepsilon/M}\right) = \left\{V \in \Gamma\left(\mathtt{A}'_\varepsilon, N_{\mathtt{A}'_\varepsilon/M}\right) | V \in C^{1,\alpha},\ V|_{\partial \mathtt{A}'_\varepsilon} \subset TC_0 \cup TC_\varepsilon\right\},$$

(the "$-$" in $C^{1,\alpha}_-$ is for the coassociative boundary condition), and

$$F_\varepsilon : C^{1,\alpha}_-\left(\mathtt{A}'_\varepsilon, N_{\mathtt{A}'_\varepsilon/M}\right) \to C^\alpha\left(\mathtt{A}'_\varepsilon, N_{\mathtt{A}'_\varepsilon/M}\right),$$

$$F_\varepsilon(V) = *_{\mathtt{A}'_\varepsilon} \circ \perp_{\mathtt{A}'_\varepsilon} \circ \left(T_V \circ (\exp V)^* \tau\right),$$

where

(1) $(\exp V)^*$ pulls back the *differential form part* of the tensor τ
(2) $T_V : T_{\exp_p(tV)}M \to T_pM$ pulls back the *vector part* of τ by parallel transport along the geodesic $\exp_p(tV)$
(3) $\perp_{\mathtt{A}'_\varepsilon} : TM|_{\mathtt{A}'_\varepsilon} \to N_{\mathtt{A}'_\varepsilon/M}$ is the orthogonal projection with respect to g
(4) $*_{\mathtt{A}'_\varepsilon} : \Omega^3(\mathtt{A}'_\varepsilon) \to \Omega^0(\mathtt{A}'_\varepsilon)$ is the quotient by the volume form $dvol_{\mathtt{A}'_\varepsilon}$

If $\mathtt{A}'_\varepsilon$ is an almost instanton, then a G_2-linear algebra argument shows that when $\|V\|_{C^{1,\alpha}\left(A'_\varepsilon, N_{A'_\varepsilon/M}\right)}$ is small we have

$$F_\varepsilon(V) = 0 \iff \mathtt{A}_\varepsilon(V) \text{ is an instanton.}$$

To ensure that $\mathtt{A}_\varepsilon(V)$ satisfies the boundary condition, in the definition of $\exp V$ we actually need to modify the metric g near $C_0 \cup C_\varepsilon$ to make them totally geodesic, but we will keep the original metric in T_V, $\perp_{\mathtt{A}'_\varepsilon}$ and $*_{\mathtt{A}'_\varepsilon}$. This modification will not change the expression of $F'_\varepsilon(0)$ in (7) (see Remark 10 (1) in [21]). So our estimate for $\left\|F'_\varepsilon(0)^{-1}\right\|$ is still valid in the new metric.

3.2.2. Linearizing the instanton equation using a good frame

We can make $F_\varepsilon(V)$ more explicit by using a good local frame field $\{W_\alpha\}_{\alpha=1}^7$ near $p \in A'_\varepsilon = \mathtt{A}_\varepsilon(0)$. Since τ is a vector-valued 3-form, following Einstein's summation convention, we can write

$$\tau = \omega^\alpha \otimes W_\alpha \in \Omega^3(M, TM),$$

where local 3-forms $\{\omega^\alpha\}_{\alpha=1}^7$ are determined by $\{W_\alpha\}_{\alpha=1}^7$. For any $q = \exp_p V \in \mathtt{A}_\varepsilon(V)$, we let

$$F(V)(p) = (\exp V)^* \omega^\alpha(p) \otimes T_V W_\alpha(q).$$

$F(V)$ is the essential part of $F_\varepsilon(V)$ as $*_{\mathtt{A}'_\varepsilon} \circ \perp_{\mathtt{A}'_\varepsilon}$ is only a projection. It is *independent* of the choice of frame $\{W_\alpha\}_{\alpha=1}^7$. To compute $F'(0)$, we have the following *generalized Cartan formula* (6):

$$\begin{aligned}
&F'(0)V(p) \\
&= L_V\omega^\alpha \otimes W_\alpha + \omega^\alpha \otimes \nabla_V W_\alpha \\
&= \underbrace{d(i_V\omega^\alpha) \otimes W_\alpha}_{\text{symbol part}} + \underbrace{i_V d\omega^\alpha(p) \otimes W_\alpha + \omega^\alpha \otimes \nabla_V W_\alpha(p)}_{\text{0-th order part}}. \qquad (6)\\
&= d(i_V\omega^\alpha) \otimes W_\alpha(p) + B_\alpha(p)V \otimes W_\alpha + \omega^\alpha \otimes C_\alpha(p)V \\
&= \mathcal{D}V \otimes vol_{A'_\varepsilon} + E(p)(V) \qquad (7)
\end{aligned}$$

where B_α, C_α and E are certain matrix-valued functions.

We require a *"good"* frame field $\{W_\alpha\}_{\alpha=1}^7$ to satisfy the following conditions in any small *ε-ball* around *p:*

(1) B_α, C_α and E are of ε-order in C^1 norm,
(2) ω^a's are ε-close to ω^{ijk}'s in (3) for $\mathbb{R}^7$ in C^1 norm, in the sense that ω^{ijk} are replaced by $W_i^* \wedge W_j^* \wedge W_k^*$, where W_i^* is the dual vector of W_i,
(3) $\mathcal{D}V(p)$ is ε-close to the twisted Dirac operator (4) for $\mathbb{R}^7$ in C^1 norm, in the sense that $\frac{\partial}{\partial x^i}$ in (4) are replaced by $\nabla^\perp_{W_i}$.

Condition 1 holds when $\{W_\alpha\}_{\alpha=1}^7$ is parallel along the normal bundle directions. Conditions 2 holds if whenever $e_\gamma = e_\alpha \times e_\beta$ in $\operatorname{Im}\mathbb{O}$ we have $\|W_\gamma - W_\alpha \times W_\beta\|_{C^1} = O(\varepsilon)$. Condition 3 holds if we further have the *normal covariant derivatives* $\|\nabla^\perp_{W_i} W_k\|_{C^1} = O(\varepsilon)$ assuming that W_i's span TA'_ε and W_k's span $N_{A'_\varepsilon/M}$.

Such a *good* frame $\{W_\alpha\}_{\alpha=1}^7$ can be constructed by the Cayley-Dickson construction as explained in Remark 3. The principal part of the linearized instanton equation $F'(0)V$ on A'_ε is the term $\mathcal{D}V$ in (7), which is a first order differential operator with a nice geometric meaning (see the next subsection).

3.2.3. A simplified model: Dirac operators on thin manifolds

We temporarily leave G_2 geometry and consider a (*"thin"* when ε is small) 3-manifold

$$\mathtt{A}_\varepsilon := [0,\varepsilon] \times \Sigma = \{(x_1, z := x_2 + ix_3)\},$$

with the *warped product metric*

$$g_{\mathtt{A}_\varepsilon,h} = h(z)\, dx_1^2 + g_\Sigma,$$

where $h(z) = |n|^2 > 0$ and $n = \frac{dC_t}{dt}|_{t=0}$ is the nonvanishing normal vector field on C_0. This is a first order approximation of the induced metric on $\mathtt{A}'_\varepsilon \subset M$.

We first consider the geometry of a J_n-holomorphic curve Σ in C. Let $L = N_{\Sigma/C}$ be the normal bundle of Σ in C, then L is a Hermitian J_n-complex line bundle over Σ (see Proposition 24). Let $\bar\partial = \left(\overline{\partial}, \overline{\partial}^*\right)$ be the Dirac operator on the Dolbeault complex $\Omega^0_\mathbb{C}(L) \oplus \Omega^{0,1}_\mathbb{C}(L)$ of the spinor bundle of Σ

$$\mathbb{S}_\Sigma = \mathbb{S}^+_\Sigma \oplus \mathbb{S}^-_\Sigma = L \oplus \wedge^{0,1}_\mathbb{C}(L)$$

such that

$$\Omega^0_\mathbb{C}(L) \oplus \Omega^{0,1}_\mathbb{C}(L) \overset{(\bar\partial,\bar\partial^*)}{\longrightarrow} \Omega^{0,1}_\mathbb{C}(L) \oplus \Omega^0_\mathbb{C}(L), \tag{8}$$

where $\overline{\partial} : \Omega^0_\mathbb{C}(L) \to \Omega^{0,1}_\mathbb{C}(L)$ is the *normal Cauchy-Riemann operator* of J_n-holomorphic curve Σ in C, and $\overline{\partial}^* : \Omega^{0,1}_\mathbb{C}(L) \to \Omega^0_\mathbb{C}(L)$ is its adjoint. (In [21], we use the notation $\partial^+ = -\,\mathbf{i}\overline{\partial}^*$ and $\partial^- = \mathbf{i}\overline{\partial}$).

To describe the spinor bundle $\mathbb{S}$ over the 3-manifold $\mathtt{A}_\varepsilon$, we pullback $\mathbb{S}_\Sigma = \mathbb{S}^+_\Sigma \oplus \mathbb{S}^-_\Sigma$ to $\mathtt{A}_\varepsilon = [0,\varepsilon] \times \Sigma$, and denote it as $\mathbb{S} = \mathbb{S}^+ \oplus \mathbb{S}^-$. Let e_1 be the unit tangent vector field on $\mathtt{A}_\varepsilon$ along x_1-direction. For $\mathbb{S}$ to be the spinor bundle of $\mathtt{A}_\varepsilon$, the Clifford multiplication with e_1 should be $\pm i$ on $\mathbb{S}^\pm$. To describe the Dirac operator on $\mathbb{S}$, we define

$$\mathcal{D} = e_1 \cdot h^{-\frac{1}{2}}(z) \frac{\partial}{\partial x_1} + \bar\partial,$$

where $\bar\partial = \left(\overline{\partial}, \overline{\partial}^*\right)$ is the Dirac Dolbeault operator in equation (8). $\mathcal{D}$ acts on the sections $V = (u,v)$ of the spinor bundle $\mathbb{S} = \mathbb{S}^+ \oplus \mathbb{S}^-$ over $\mathtt{A}_\varepsilon$ with *local elliptic boundary condition* (in the sense of [5]):

$$v|_{\partial \mathtt{A}_\varepsilon} = 0.$$

We can write a local expression for $\mathcal{D}$. Consider the section $V = (u,v)$ of $\mathbb{S}$ with $u = V^4 + \mathbf{i}V^5 \in \mathbb{S}^+$ and $v = V^6 + \mathbf{i}V^7 \in \mathbb{S}^-$ (for $\mathbb{S}^\pm$ are complex line bundles). Then

$e_1 \cdot = \begin{bmatrix} \mathbf{i} & 0 \\ 0 & -\mathbf{i} \end{bmatrix}$ and

$$\begin{aligned} \mathcal{D}V &= \begin{bmatrix} \mathbf{i} & 0 \\ 0 & -\mathbf{i} \end{bmatrix} \left(h^{-\frac{1}{2}}(z) \frac{\partial}{\partial x_1} + \begin{bmatrix} 0 & \mathbf{i}\partial_z \\ \mathbf{i}\bar{\partial}_z & 0 \end{bmatrix} \right) \begin{bmatrix} u \\ v \end{bmatrix}, \\ &= \left(h^{-\frac{1}{2}}(z) \frac{\partial u}{\partial x_1} \mathbf{i} - \partial_z v \right) + \left(-h^{-\frac{1}{2}}(z) \frac{\partial v}{\partial x_1} \mathbf{i} + \bar{\partial}_z u \right) \cdot \mathbf{j} \end{aligned} \tag{9}$$

where $\bar{\partial}_z := \nabla_2 + \mathbf{i}\nabla_3$ and $\partial_z := \nabla_2 - \mathbf{i}\nabla_3$.

One can check that $\mathcal{D}$ agrees with the linearized instanton equation on $\{0\} \times \Sigma$, and on A_ε they are very close (Subsection 3.2.6). This is why we can use $\mathcal{D}$ of the linear model to study deformations of instantons in G_2 manifolds. The precise comparison is in Subsection 3.2.5 and Subsection 3.2.6.

3.2.4. Key estimates of $\mathcal{D}^{-1}$ of the linear model

The most difficult part of [21] is to derive an explicit ε-dependent bound of the operator norm of $\mathcal{D}^{-1} : C^\alpha(\mathtt{A}_\varepsilon, \mathbb{S}) \to C_-^{1,\alpha}(\mathtt{A}_\varepsilon, \mathbb{S})$. This is nontrivial because we lose uniform ellipticity as $\mathtt{A}_\varepsilon = [0,\varepsilon] \times \Sigma$ collapses to Σ and consequently $\|\mathcal{D}^{-1}\|$ may blow up. A good control of $\|\mathcal{D}^{-1}\|$ is crucial for singular perturbation problems in general. We could achieve this in our case roughly because of

(1) The Fredholm regularity property of the J_n-holomorphic curve Σ, which supplies the transversality for $\mathcal{D}$,
(2) The coassociative boundary condition on $\partial\mathtt{A}_\varepsilon$, which enables us to periodically reflect $\mathtt{A}_\varepsilon$ to a bigger domain $\mathtt{A}_{k(\varepsilon)\varepsilon}$ with integer $k(\varepsilon)$ such that $1/2 \le k(\varepsilon)\varepsilon \le 3/2$, thus restoring the uniform ellipticity.

This is an over-simplified description as sections on $\mathtt{A}_\varepsilon$ may become *discontinuous* after periodical reflection, so condition 2 only helps in the $W^{1,p}$ setting to get a uniform bound of $\|\mathcal{D}^{-1}\|$ as in [8]. More effort is needed to estimate $\|\mathcal{D}^{-1}\|$ in the Schauder $C^{1,\alpha}$ setting.

Recall in Subsection 3.2.1 we have formulated instantons nearby A'_ε as solutions of the nonlinear equation $F_\varepsilon(V) = 0$, and it turns out $\|F_\varepsilon(0)\|_{C^\alpha} \le C\varepsilon^{1-\alpha}$. In Subsection 3.2.2 we have computed the linearization $F'_\varepsilon(0)$. To apply the implicit function theorem to perturb A'_ε to a true instanton, we need to estimate $\left\|F'_\varepsilon(0)^{-1}\right\|$, and we will see in Subsection 3.2.6 it is comparable to $\|\mathcal{D}^{-1}\|$ (Proposition 25). So a key estimate that we need is the following

Theorem 15. *(ε-dependent bound) Suppose that the first eigenvalues for $\overline{\partial}\overline{\partial}^*$ and $\overline{\partial}^*\overline{\partial}$ are bounded below by $\lambda > 0$. Then for any $0 < \alpha < 1$ and $p > 3$ there is a positive constant $C = C(\alpha, p, \lambda, h)$ such that for any $V \in C_-^{1,\alpha}(\mathtt{A}_\varepsilon, \mathbb{S})$, we have*

$$\|V\|_{C_-^{1,\alpha}} \le C\varepsilon^{-\left(\frac{3}{p}+2\alpha\right)} \|\mathcal{D}V\|_{C^\alpha}. \tag{10}$$

Here the notations $C_{\pm}^{1,\alpha}(\mathtt{A}_\varepsilon,\mathbb{S})$ are chosen to indicate the *local elliptic boundary conditions* for sections $V=(u,v)$ of $\mathbb{S}=\mathbb{S}^+\oplus\mathbb{S}^-$:

$$C_{-}^{1,\alpha}(\mathtt{A}_\varepsilon,\mathbb{S})=\left\{V\in C^{1,\alpha}(\mathtt{A}_\varepsilon,\mathbb{S}):v|_{\partial\mathtt{A}_\varepsilon}=0\right\}$$
$$C_{+}^{1,\alpha}(\mathtt{A}_\varepsilon,\mathbb{S})=\left\{V\in C^{1,\alpha}(\mathtt{A}_\varepsilon,\mathbb{S}):u|_{\partial\mathtt{A}_\varepsilon}=0\right\}.$$

Similarly $W_{\pm}^{k,p}(\mathtt{A}_\varepsilon,\mathbb{S})$ will be used to indicate these boundary conditions.

To prove this theorem we first recall (for simplicity, we assume $h\equiv 1$),

$$\mathcal{D}V=\begin{bmatrix}\mathbf{i}&0\\0&-\mathbf{i}\end{bmatrix}\left(\frac{\partial}{\partial x_1}+\begin{bmatrix}0&-\mathbf{i}\overline{\partial}^*\\\mathbf{i}\overline{\partial}&0\end{bmatrix}\right)\begin{bmatrix}u\\v\end{bmatrix}. \tag{11}$$

Along Σ we have the following

Lemma 16. *Assume the J_n-holomorphic curve $\Sigma\subset C$ is Fredholm regular, then*

$$\lambda_{\overline{\partial}^*}>0$$

where $\lambda_{\overline{\partial}^*}$ is the first eigenvalue of $\Delta_\Sigma=\overline{\partial}\overline{\partial}^*$ on $W^{1,2}(\Sigma,\mathbb{S}^+)$.

Proof. This follows from the fact that $\overline{\partial}$ is the normal Cauchy-Riemann operator on $N_{\Sigma/C}$, and its adjoint is $\overline{\partial}^*$. So the Fredholm regular property of Σ is equivalent to $\ker\overline{\partial}^*=\{0\}$, i.e., $\lambda_{\overline{\partial}^*}>0$. □

L^2 estimate

The operator $\mathcal{D}=\mathcal{D}_\pm : W_{\pm}^{1,2}(\mathrm{A}_\varepsilon,\mathbb{S})\to L^2(\mathrm{A}_\varepsilon,\mathbb{S})$ is self-adjoint by the boundary condition, thus coker$\mathcal{D}_-=\ker\mathcal{D}_+$. By the Rayleigh quotient method, we know

$$\lambda_{\mathcal{D}_\pm}:=\inf_{0\neq V\in W_{\pm}^{1,2}(\mathtt{A}_\varepsilon)}\frac{\|\mathcal{D}V\|^2_{L^2(\mathtt{A}_\varepsilon)}}{\|V\|^2_{L^2(\mathtt{A}_\varepsilon)}} \tag{12}$$

is the first eigenvalue of the Laplacian $\mathcal{D}_\mp\mathcal{D}_\pm$.

Theorem 17 (First eigenvalue estimate)**.** *For $\mathcal{D}_\pm : W_{\pm}^{1,2}(\mathtt{A}_\varepsilon,\mathbb{S})\to L^2(\mathtt{A}_\varepsilon,\mathbb{S})$, we have*

$$\lambda_{\mathcal{D}_-}\geq\min\left\{\lambda_{\overline{\partial}},\frac{2}{\varepsilon^2}\right\}\text{ and }\lambda_{\mathcal{D}_+}\geq\min\left\{\lambda_{\overline{\partial}^*},\frac{2}{\varepsilon^2}\right\}.$$

Remark 18. *Theorem 17 enables us to control the first eigenvalue of $\mathcal{D}_\pm$ on 3-dimensional $\mathtt{A}_\varepsilon$ by that of $\overline{\partial}$ on 2-dimensional Σ, when ε is small. The control is due to the boundary condition of $\mathcal{D}_\pm$, as will be clear from the following proof.*

Proof. We prove the estimate for $\lambda_{\mathcal{D}_-}$ ($\lambda_{\mathcal{D}_+}$ is similar). By the boundary condition $v|_{\partial \mathtt{A}_\varepsilon} = 0$ we have

$$\begin{aligned}\langle \mathcal{D}V, \mathcal{D}V\rangle_{L^2} &= \int_{[0,\varepsilon]\times\Sigma} \left(\left|\frac{\partial V}{\partial x_1}\right|^2 + \left\|\overline{\partial}^* v\right\|^2 + \left\|\overline{\partial} u\right\|^2 \right) \\ &\geq \int_{[0,\varepsilon]\times\Sigma} \left(\left\|\overline{\partial} u\right\|^2 + \|v_{x_1}\|^2 \right).\end{aligned}$$

Then use the Rayleigh quotient for $\overline{\partial}^*\overline{\partial}$ and notice that $v|_{\partial \mathtt{A}_\varepsilon} = 0$. □

Corollary 19. *If $\Sigma \subset C$ is Fredholm regular, then for small enough $\varepsilon > 0$, the map $\mathcal{D} : W^{1,2}_- (\mathtt{A}_\varepsilon, \mathbb{S}) \to L^2 (\mathtt{A}_\varepsilon, \mathbb{S})$ is surjective.*

Proof. By Lemma 16, the Fredholm regular property of Σ implies that $\lambda_{\overline{\partial}^*} > 0$. So by Theorem 17, $\lambda_{\mathcal{D}_+} > 0$, i.e. $\ker \mathcal{D}_+ = \{0\}$. By the self-adjoint property of $\mathcal{D}$, we get $\mathrm{coker}\mathcal{D}_- = \ker \mathcal{D}_+ = \{0\}$. □

Now, from the definition of $\lambda_{\mathcal{D}_-}$ we obtain the L^2 estimate

Corollary 20 (L^2-estimate)**.** *For any $V \in W^{1,2}_- (\mathtt{A}_\varepsilon, \mathbb{S})$ and small enough ε, we have*

$$\|V\|_{W^{1,2}_-(\mathtt{A}_\varepsilon,\mathbb{S})} \leq C(\lambda) \|\mathcal{D}V\|_{L^2(\mathtt{A}_\varepsilon,\mathbb{S})}, \tag{13}$$

where the constant $C(\lambda)$ only depends on $\lambda_{\overline{\partial}}$.

In the following we will derive a C^0-estimate of V. For this purpose let us examine the equation $\mathcal{D}_- V = W$ more closely.

For $V = (u, v)$, $W = (w_1, w_2)$, the equation $\mathcal{D}_- V = W$ (assuming $h = 1$) is equivalent to the following system

$$\left\{ \begin{array}{rcl} u_{x_1} - \mathbf{i}\overline{\partial}^* v &=& w_1 \\ v_{x_1} + \mathbf{i}\overline{\partial} u &=& w_2 \end{array} \right. \quad \text{and} \quad v|_{\partial \mathtt{A}_\varepsilon} = 0. \tag{14}$$

This is similar to the Cauchy-Riemann equation, but the relation between u and v is *weaker*, since ∇u can only control $\overline{\partial}^* v$ and v_{x_1}, which are only *half* of partial derivatives of v; the same applies to ∇v. This issue makes it more difficult to obtain the C^0-estimate of u than in the Cauchy-Riemann type equations.

C^0 estimate

The C^0-estimate of V is derived from a $W^{1,p}$-estimate of V and Sobolev embedding. To obtain a uniform estimate for $\left\|\mathcal{D}^{-1}\right\|$ in the L^p setting, we use the periodic reflection technique.

Theorem 21 (L^p-estimate, $p>3$). *For any* $V \in W_{-}^{1,p}(\mathtt{A}_\varepsilon, \mathbb{S})$, *we have*

$$\|V\|_{W_{-}^{1,p}(\mathtt{A}_\varepsilon,\mathbb{S})} \le C_p(\lambda)\|\mathcal{D}V\|_{L^p(\mathtt{A}_\varepsilon,\mathbb{S})}, \tag{15}$$

where the constant $C_p(\lambda)$ *only depends on* $\lambda_{\overline{\partial}}$, $Vol(\Sigma)$, *and* p.

Proof. Given any $\varepsilon>0$, we choose an integer $k(\varepsilon)$ so that $1/2 \le k(\varepsilon)\varepsilon \le 3/2$. In the following we will simply write $k(\varepsilon)$ as k. We reflect A_ε to $\mathrm{A}_{k\varepsilon}$ periodically and extend (u,v) along the boundaries $\varepsilon\mathbb{Z}\times\Sigma$ such that

$$v(x,z) = \begin{cases} -v((2j+2)\varepsilon - x, z) & x \in [(2j+1)\varepsilon, (2j+2)\varepsilon] \\ v(x-2j\varepsilon, z) & x \in [2j\varepsilon, (2j+1)\varepsilon] \end{cases},$$

$$u(x,z) = \begin{cases} u((2j+2)\varepsilon - x, z) & x \in [(2j+1)\varepsilon, (2j+2)\varepsilon] \\ u(x-2j\varepsilon, z) & x \in [2j\varepsilon, (2j+1)\varepsilon] \end{cases},$$

(Notice that a $W^{k,p}$ section will remain so after reflections), i.e., we do *odd extensions* of v and *even extensions* of u along the boundaries. By (14), this induces *odd extensions* of w_1 and *even extensions* of w_2 along the boundaries. Since the shape of $\mathrm{A}_{k\varepsilon}$ is uniformly bounded, we have the elliptic estimate

$$\tilde{C}(p)\|V\|_{W_{-}^{1,p}(\mathtt{A}_{k\varepsilon},\mathbb{S})} \le \|\mathcal{D}V\|_{L^p(\mathtt{A}_{2k\varepsilon}\cup\mathtt{A}_{-2k\varepsilon},\mathbb{S})} + \|V\|_{L_{-}^{p}(\mathtt{A}_{2k\varepsilon}\cup\mathtt{A}_{-2k\varepsilon},\mathbb{S})},$$

with a uniform constant $\tilde{C}(p)$ for all ε. Then we use the interpolation inequality

$$\|V\|_{L_{-}^{p}(\mathtt{A}_{k\varepsilon}\cup\mathtt{A}_{-2k\varepsilon},\mathbb{S})} \le C\left(\delta^{\frac{p}{p-1}}\|V\|_{W^{1,p}(\mathtt{A}_{k\varepsilon}\cup\mathtt{A}_{-2k\varepsilon},\mathbb{S})} + \delta^{-p}\|V\|_{L_{-}^{2}(\mathtt{A}_{k\varepsilon}\cup\mathtt{A}_{-2k\varepsilon},\mathbb{S})}\right) \tag{16}$$

to pass from the L^2 estimate to the L^p estimate

$$\|V\|_{W_{-}^{1,p}(\mathtt{A}_{k\varepsilon},\mathbb{S})} \le C_p(\lambda)\|\mathcal{D}V\|_{L^p(\mathtt{A}_{k\varepsilon},\mathbb{S})}. \tag{17}$$

This is because the boundary condition of $V \in W_{-}^{1,p}(\mathtt{A}_\varepsilon,\mathbb{S})$ and (13) yields

$$\|V\|_{L_{-}^{2}(\mathtt{A}_{k\varepsilon}\cup\mathtt{A}_{-2k\varepsilon},\mathbb{S})} \le 2C(\lambda)\|\mathcal{D}V\|_{L^2(\mathtt{A}_{k\varepsilon},\mathbb{S})} \le C_p(\lambda)\|\mathcal{D}V\|_{L^p(\mathtt{A}_{k\varepsilon},\mathbb{S})},$$

where $C_p(\lambda) = 2C(\lambda)\left(\frac{3}{2}Vol(\Sigma)\right)^{\frac{1}{2}-\frac{1}{p}}$. Last we obtain the inequality (15) on $\mathtt{A}_\varepsilon$ from (17) by the periodicity of the L^p integrals on the reflected domains. □

Corollary 22 (C^0 estimate). *If for each* $z\in\Sigma$ *there exist* $x,x'\in[0,\varepsilon]$ *such that both* $u(x,z)=0$ *and* $v(x',z)=0$, *then*

$$\|V\|_{C^0(\mathtt{A}_\varepsilon,\mathbb{S})} \le C\varepsilon^{1-\frac{3}{p}}\|\mathcal{D}V\|_{C^0(\mathtt{A}_\varepsilon,\mathbb{S})}. \tag{18}$$

Proof. We have

$$\begin{aligned} \|V\|_{C^0(\mathtt{A}_{k\varepsilon},\mathbb{S})} &\overset{\text{Schauder}}{\le} C\varepsilon^{1-\frac{3}{p}}\|V\|_{C^{1-\frac{3}{p}}(\mathtt{A}_{k\varepsilon},\mathbb{S})} \overset{\text{Sobolev}}{\le} C\varepsilon^{1-\frac{3}{p}}\|V\|_{W^{1,p}(\mathtt{A}_{k\varepsilon},\mathbb{S})} \\ &\le C\varepsilon^{1-\frac{3}{p}}\|\mathcal{D}V\|_{L^p(\mathtt{A}_{k\varepsilon},\mathbb{S})}\ (\text{by } (15)) \le C\varepsilon^{1-\frac{3}{p}}\|\mathcal{D}V\|_{C^0(\mathtt{A}_{k\varepsilon},\mathbb{S})}, \end{aligned}$$

where in the first inequality we have used the condition $u(x,z)=0$, and $v(x',z)=0$ and the definition of the Schauder $C^{1-\frac{3}{p}}$ norm. Then we notice the periodicity on $\mathtt{A}_{k\varepsilon}$ so we get (18). □

$C^{1,\alpha}$-estimate

We can change $\mathcal{D}_{-}V=W$ (14) into a system of second order elliptic equations

$$\begin{cases} u_{x_1x_1}-\overline{\partial}^*\overline{\partial}u=\mathbf{i}\overline{\partial}^*w_2+\partial_{x_1}w_1 \\ v_{x_1x_1}-\overline{\partial}\,\overline{\partial}^*v=-\mathbf{i}\overline{\partial}w_1+\partial_{x_1}w_2 \end{cases} \quad \text{and} \quad v|_{\partial\mathtt{A}_\varepsilon}=0, \tag{19}$$

where $V=(u,v)$ and $W=(w_1,w_2)$.

To get the uniform estimate of $\left\|\mathcal{D}^{-1}\right\|$ in the $C^{1,\alpha}$ setting, we can not just rely on periodic reflections, because if $w_1\neq 0$ then the extension of $\mathcal{D}V$ is no longer continuous on $\mathtt{A}_{k\varepsilon}$, not to mention in $C^\alpha(\mathtt{A}_{k\varepsilon},\mathbb{S})$.

Theorem 23 ($C^{1,\alpha}$ estimate). *For any $V\in C_{-}^{1,\alpha}(\mathtt{A}_\varepsilon,\mathbb{S})$, we have*

$$C\varepsilon^{\frac{3}{p}+2\alpha}\|V\|_{C_{-}^{1,\alpha}(\mathtt{A}_\varepsilon,\mathbb{S})}\leq\|\mathcal{D}V\|_{C^\alpha(\mathtt{A}_\varepsilon,\mathbb{S})}.$$

Proof. We can decouple $W=(w_1,w_2)$ into dealing with $(0,w_2)$ and $(w_1,0)$ cases separately because $\mathcal{D}$ is linear and *surjective* from Corollary 19.

When $w_1=0$, along $\partial\mathrm{A}_\varepsilon$ we have $u_{x_1}=\overline{\partial}^*v=0$ (for $v|_{\partial\mathtt{A}_\varepsilon}=0$), so we can reflect u by even extension and it is still in $C^{1,\alpha}$. We can extend W in $C^\alpha(\mathtt{A}_\varepsilon,\mathbb{S})$ as well, for $w_1=0$. In this case we can *restore uniform ellipticity* in the Schauder setting by periodic reflection.

We deal with the more difficult case when $w_1\neq 0$ but $w_2=0$. The v component is easier, for it satisfies the Dirichlet boundary condition $v|_{\partial\mathrm{A}_\varepsilon}=0$. Standard Schauder estimate on half balls ([11]) implies

$$C\varepsilon^{1+\alpha}\|v\|_{C_{-}^{1,\alpha}(\mathtt{A}_\varepsilon,\mathbb{S})}\leq\varepsilon\|w_1\|_{C^\alpha(\mathtt{A}_\varepsilon,\mathbb{S}+)}+\|v\|_{C_{-}^{0}(\mathtt{A}_\varepsilon,\mathbb{S})}. \tag{20}$$

Plugging the C^0 estimate (18) in above inequality we get

$$\|v\|_{C_{-}^{1,\alpha}(\mathtt{A}_\varepsilon,\mathbb{S})}\leq C\varepsilon^{-\left(\frac{3}{p}+\alpha\right)}\|W\|_{C^\alpha(\mathtt{A}_\varepsilon,\mathbb{S})}.$$

The u component is much harder, since $u|_{\partial\mathrm{A}_\varepsilon}\neq 0$ in general. We carry out the following steps:

(a) We *homogenize* u by introducing

$$\widetilde{u}=u-\rho\left(\frac{x_1}{\varepsilon}\right)u(\varepsilon,z)-\left(1-\rho\left(\frac{x_1}{\varepsilon}\right)\right)u(0,z),$$

where $\rho:[0,1]\to[0,1],\rho(0)=0,\rho(1)=1$ is a smooth cut-off function such that $\|\rho\|_{C^{2,\alpha}[0,1]}\leq C$. Then we get

$$\widetilde{u}_{x_1x_1}-\partial^+\partial^-\widetilde{u}=\partial_{x_1}\widetilde{w_1}+g \quad \text{and} \quad \widetilde{u}|_{\partial\mathtt{A}_\varepsilon}=0,$$

where the functions $\widetilde{w_1}$ and g involve w_1 and the boundary values of u. Applying elliptic regularity for this Dirichlet problem of $\widetilde{u}$ and then writing back in u, we have

$$\varepsilon \|w_1\|_{C^\alpha} + \varepsilon^{1+\alpha} \|u\|_{C^0_-} + \|v\|_{C^0_-} \geq C\varepsilon^{1+2\alpha} \|u\|_{C^{1,\alpha}_-}. \tag{21}$$

This is not quite easy, since we can not use ∇v to completely control ∇u from (14). This was done by carefully analyzing the ε-orders behaviors of the derivatives of the cut-off function $\rho\left(\frac{x_1}{\varepsilon}\right)$ and elliptic estimates of $\overline{\partial}$ and $\overline{\partial}^*$ on slices $\{x\} \times \Sigma$.

(b) We show on each segment $[0,\varepsilon] \times \{z\}$ *there exists* x *with* $u(x,z) = 0$. Consider the *average* of u along each segment $[0,\varepsilon] \times \{z\}$:

$$\overline{u}(z) = \int_0^\varepsilon u(x_1, z)\, dx_1.$$

From (14) we have (note that $w_2 = 0$)

$$\mathbf{i}\overline{\partial}\overline{u}(z) = -\int_0^\varepsilon \partial_{x_1} v(x_1, z)\, dx_1 = v(0,z) - v(\varepsilon, z) = 0.$$

But from the assumption that $\overline{\partial}^*\overline{\partial}$ has trivial kernel on Σ we get that

$$\overline{u}(z) \equiv 0.$$

Then by the intermediate value theorem there must exist some $x \in [0,\varepsilon]$ such that $u(x,z) = 0$. So by Corollary 22 we get the desired C^0 estimate of u. Similarly we get an estimate for v.

(c) Putting the C^0 estimate of $V = (u,v)$ back to (21), we finally get

$$\|V\|_{C^{1,\alpha}_-(\mathtt{A}_\varepsilon, \mathbb{S})} \leq C\varepsilon^{-\left(\frac{3}{p}+2\alpha\right)} \|DV\|_{C^\alpha(\mathtt{A}_\varepsilon, \mathbb{S})}.$$

Theorem 15 is proved. □

3.2.5. Geometry of J_n-holomorphic curves in G_2 manifolds: agreement of two Dirac operators

Recall given a family $\mathcal{C} = \cup_{0\leq t\leq\varepsilon} C_t$ of coassociative manifolds C_t in M, the (nonvanishing) deformation vector field $n := \frac{dC_t}{dt}|_{t=0}$ defines an almost complex structure J_n on $C = C_0$. For any J_n-holomorphic curve $\Sigma \subset C$, the G_2-structure on M gives a natural identification between $N_{\Sigma/C}$ and $N_{C/M}|_\Sigma$ ([10]). Further calculations [21] established the following close relation between the "intrinsic" and "extrinsic" Dirac operators.

Proposition 24 (Proposition 17, 18 in [21]). *Let* $\Sigma \subset C \subset \mathcal{C} \subset M$ *be as above. We have an orthogonal decomposition*

$$TM|_\Sigma = T\Sigma \oplus N_{\Sigma/C} \oplus N_{C/M}|_\Sigma \oplus N_{C/\mathcal{C}}|_\Sigma,$$

(1) *For* $L = T\Sigma$, $N_{\Sigma/C}$ *or* $N_{C/M}|_\Sigma$, *the induced connection* ∇^L *from the Levi-Civita connection on* M *is Hermitian, i.e.,* $\nabla^L J_n = 0$.

(2) *The spinor bundle* $\mathbb{S}_\Sigma$ *over* Σ *is identified with* $N_{\Sigma/C} \oplus N_{C/M}|_\Sigma$ *in such a way that the Clifford multiplication is given by the* G_2 *multiplication* $\times$ *and the spinor connection equals* $\nabla^{N_{\Sigma/C} \oplus N_{C/M}|_\Sigma}$.

(3) *The Dirac operator on* $\mathbb{S}_\Sigma = N_{\Sigma/C} \oplus N_{C/M}|_\Sigma$ *agrees with the Dolbeault operator on* $N_{\Sigma/C} \oplus \wedge^{0,1}_{\mathbb{C}} (N_{\Sigma/C})$.

The agreement of the two Dirac type operators on Σ is the geometric reason that the Fredholm regularity property of the "intrinsic" Cauchy-Riemann operator on $N_{\Sigma/C}$ gives control of the "extrinsic" linearized instanton operator.

3.2.6. Comparison of $\mathcal{D}$ and linearized instanton equation

When we move from Σ to the interior of the almost instanton A'_ε, the nice agreement of the two Dirac operators no longer holds. We need to control the difference between the Dirac operator $\mathcal{D}$ on $\mathbb{S}$ and the linearized instanton operator $F'_\varepsilon(0)$ on $N_{\mathtt{A}'_\varepsilon/M}$. In order to compare them, we need a *good* identification between $\mathbb{S}$ and $N_{\mathtt{A}'_\varepsilon/M}$.

For this purpose we defined an *exponential-like map* $\widetilde{\exp} : \mathbb{S} \to M$ which satisfies $\widetilde{\exp}(\mathtt{A}_\varepsilon) = \mathtt{A}'_\varepsilon$ and its differential $d\widetilde{\exp}|_{\mathtt{A}_\varepsilon}$ has the following properties on $\{0\} \times \Sigma$ (see Appendix of [21]):

(1) $d\widetilde{\exp}|_{\{0\}\times\Sigma} = (id, f) : N_{\Sigma/C} \oplus \wedge^{0,1}_{\mathbb{C}} (N_{\Sigma/C}) \to N_{\Sigma/C} \oplus N_{C/M}|_\Sigma$ on fiber directions of $\mathbb{S}$, where f is the isomorphism between the second summands given in Lemma 3.2 of [10];

(2) $d\widetilde{\exp}|_{\{0\}\times\Sigma} = id : T\Sigma \to T\Sigma$ and $d\widetilde{\exp}|_{\{0\}\times\Sigma} : \frac{\partial}{\partial x_1} \to n(z)$ on base directions of $\mathbb{S}$;

(3) $\widetilde{\exp}|_{\{0\}\times\Sigma}(\mathbb{S}^+ \oplus 0) \subset C_0$, $\widetilde{\exp}|_{\{\varepsilon\}\times\Sigma}(\mathbb{S}^+ \oplus 0) \subset C_\varepsilon$ on the boundary $\{0, \varepsilon\} \times \Sigma$ of $\mathtt{A}_\varepsilon$.

Using the map $d\widetilde{\exp}|_{\mathtt{A}_\varepsilon}$, we can relate the *spin bundle* $\mathbb{S} \to \mathtt{A}_\varepsilon$ to the *normal bundle* $N_{\mathtt{A}'_\varepsilon/M} \to \mathtt{A}'_\varepsilon$. Because of the compatibilities given in 1-3 above, using Proposition 24 we have the following comparison result.

Proposition 25 (Proposition 21 in [21]). *For any* $V_1 \in C^{1,\alpha}(A_\varepsilon, \mathbb{S})$, *we have*

$$\|F'_\varepsilon(0)(d\widetilde{\exp} \cdot V_1) - (d\widetilde{\exp}) \circ \mathcal{D} V_1\|_{C^\alpha(A'_\varepsilon, N_{A'_\varepsilon/M})} \leq C\varepsilon^{1-\alpha} \|V_1\|_{C^{1,\alpha}(A_\varepsilon,\mathbb{S})} \cdot$$

Then we get the uniform inverse estimate of $F'_\varepsilon(0)$ from $\|\mathcal{D}^{-1}\|$ by the following diagram, using that $d\widetilde{\exp}$ and $(d\widetilde{\exp})^{-1}$ are both smooth with a uniform C^2 bound for all ε:

$$\begin{array}{ccc}
C^{1,\alpha}_-(\mathtt{A}_\varepsilon, \mathbb{S}) & \xrightarrow{\mathcal{D}} & C^\alpha(\mathtt{A}_\varepsilon, \mathbb{S}) \\
{\scriptstyle d\widetilde{\exp}} \downarrow & & \downarrow {\scriptstyle d\widetilde{\exp}} \\
C^{1,\alpha}_-(N_{\mathtt{A}'_\varepsilon/M}) & \xrightarrow{F'_\varepsilon(0)} & C^\alpha(N_{\mathtt{A}'_\varepsilon/M})
\end{array}$$

3.2.7. Quadratic estimate

Using any local frame field $\{W_\alpha\}_{\alpha=1}^{7}$, we could compare the linearizations of $F(V)$ at two different almost instantons, up to a curvature term B as follows:

$$\begin{aligned} &F'(V_0)\,|_{A_{\varepsilon}(0)} V(p) \\ &= \left((\exp V_0)^* \otimes T_{V_0}\right) \circ \left[F'(0)\,|_{A_\varepsilon(V_0)} V_1(q)\right] + B^\alpha(V_0, V)\, W_\alpha(p), \end{aligned} \tag{22}$$

where $V_1 = (d\exp V_0)\,|_{A_\varepsilon(0)} V$. This means the derivative $F'(V_0)$ on $A_\varepsilon(0)$ can be expressed by the derivative $F'(0)$ on $A_\varepsilon(V_0)$ via the transform $(\exp V_0)^* \otimes T_{V_0}$, up to the curvature term $B^\alpha(V_0, V)\, W_\alpha(p)$. Using (22) we get

$$F'(V_0)\,V(p) - F'(0)\,V(p) = (I) + (II)$$

with

$$\begin{aligned} (I) &= \left[(\exp V_0)^* d\left(i_{V_1}\omega^\alpha\right) - d\left(i_V\omega^\alpha\right)\right]|_{A_\varepsilon(0)} \otimes W_\alpha(p), \\ (II) &= \left((\exp V_0)^* \otimes T_{V_0}\right) \circ \left[i_{V_1} d\omega^\alpha \otimes W_\alpha + \omega^\alpha \otimes \nabla_{V_1} W_\alpha\right](p) + B^\alpha(V_0, V)\, W_\alpha(p). \end{aligned}$$

Here (I) consists of 1st order terms whose C^α-norm are bounded by $\|V_0\|_{C^{1,\alpha}} \|V\|_{C^{1,\alpha}}$ because d, the pull back operator, the parallel transport and various exponential maps are Frechet smooth with respect to variations of V. (II) consists of 0th order terms and it is easier to bound. Thus we get

Proposition 26 (Quadratic estimate). *For any $V_0, V \in C^{1,\alpha}\left(A'_\varepsilon, N_{A'_\varepsilon/M}\right)$ with small $\|V_0\|_{C^{1,\alpha}}$, we have*

$$\|F'_\varepsilon(V_0)\,V - F'_\varepsilon(0)\,V\|_{C^\alpha} \le C\,\|V_0\|_{C^{1,\alpha}} \|V\|_{C^{1,\alpha}}, \tag{23}$$

where the constant C is independent of ε.

Remark 27. *We also have some pointwise estimates tied to the feature that τ is a cubic-form. In [21] we derived that*

$$\begin{aligned} &|F'(V_0)\,V - F'(0)\,V|(p) \\ &\le \left[C_7\left(|d\varphi| + |\nabla V_0|\right)^3 \left(|V_0|\,|V| + |\nabla V_0|\,|V| + |V_0|\,|\nabla V|\right) + C_5\,|V_0|\,|V|\right](p), \end{aligned}$$

where $\varphi : A_\varepsilon \to M$ is the embedding we used to define A'_ε. Because of the cubic terms, the following quadratic estimate needed in the implicit function theorem in $W^{1,p}$ setting

$$\|F'(V_0)\,V - F'(0)\,V\|_{L^p} \le C\,\|V_0\|_{W^{1,p}} \|V\|_{W^{1,p}}.$$

appears unavailable, for $\left(|\nabla V_0|^3\right)^p \notin L^p$ in general. In contrast, the Cauchy-Riemann operator of J-holomorphic curves is more linear in the L^p setting: it has the quadratic estimate (Proposition 3.5.3 in [23])

$$\|F'(V_0)\,V - F'(0)\,V\|_{L^p(\Sigma)} \le C\,\|V_0\|_{W^{1,p}(\Sigma)} \|V\|_{W^{1,p}(\Sigma)}.$$

This is one of the key reasons that we use the Schauder setting.

By our construction of the almost instanton $A'_\varepsilon = \varphi(A_\varepsilon)$ and the smoothness of φ, it is not hard to get the error estimate

$$\|F_\varepsilon(0)\|_{C^\alpha\left(A'_\varepsilon, N_{A'_\varepsilon/M}\right)} \leq C\varepsilon^{1-\alpha}.$$

Combining (10), (23), by the implicit function theorem (e.g., Proposition A.3.4 in [23]) we can solve $F_\varepsilon(V) = 0$, thus our main result, Theorem 13, is proved. QED.

4. Applications and further discussions

4.1. New examples of instantons

Our main theorem can be used to construct new examples of instantons.

Let X be a Calabi-Yau threefold containing a complex surface $S \subset X$ which contains a smooth curve $\Sigma \subset S$ that satisfies (i) $H^0(S, K_S) \neq 0$ (i.e. $p_g(S) \neq 0$) and (ii) $H^0(\Sigma, K_S|_\Sigma) = 0$. Condition (i) implies that S can be deformed inside X and condition (ii) is equivalent to $H^1\left(\Sigma, N_{\Sigma/S}\right) = 0$, namely $\Sigma \subset S$ is Fredholm regular.

Let $\{S_t\}_{0\leq t\leq\varepsilon}$ be a smooth family of deformations of S inside X and let $v = \frac{dS_t}{dt}|_{t=0}$ be the normal vector field on $S = S_0$. By our assumption, v is nontrivial, and after possible rescaling of the parameter t for the family S_t, we may assume v is very small. Let $M = X \times S^1$ be the G_2 manifold as in Example 2, and $C_t := S_t \times \{t\} \subset M$ $(0 \leq t \leq \varepsilon)$ be the family of coassociative submanifolds. They are disjoint since their second components t are different. Let $n_0 := \left(0, \frac{\partial}{\partial\theta}\right)$ be the normal vector field on $C_0 = S \times \{0\}$, then the original complex structure J_0 on S is induced from n_0, i.e., $J_0 = \frac{n_0}{|n_0|}\times$.

Let $n = \frac{dC_t}{dt}|_{t=0} = \left(v, \frac{\partial}{\partial\theta}\right)$ be the other normal vector field on C_0. Then n is nonvanishing since $\frac{\partial}{\partial\theta}$ is nonvanishing on S^1. The almost complex structure $J_n = \frac{n}{|n|}\times$ is close to but not equal to the original complex structure J_0 on S, because n is close to but not equal to n_0.

There must exist a Fredholm regular J_n-holomorphic curve $\Sigma_n \subset S$ near the original J_0-holomorphic curve Σ, because $\Sigma \subset S$ is Fredholm regular and will persist after small perturbations of J_0 on S.

Applying our main theorem to $\Sigma_n \subset C_0$, we get an instanton $A \subset M$ with boundaries on $C_0 \cup C_\varepsilon$. It is *not* the trivial instanton $\Sigma \times [0, \varepsilon]$, which has upper boundary lying on $S \times \{\varepsilon\}$, not on $C_\varepsilon = S(\varepsilon) \times \{\varepsilon\}$.

4.2. Further remarks

A few remarks on our main theorem are in order: First, counting such thin instantons is basically a problem in four manifold theory because of Bryant's result [6] which says that the zero section C in $\Lambda^2_+(C)$ is always a coassociative submanifold for an incomplete G_2-metric on its neighborhood provided that the bundle $\Lambda^2_+(C)$ is topologically trivial.

Second, when the normal vector field $n = \frac{dC_t}{dt}|_{t=0}$ has zeros, our main Theorem 13 should still hold true (work in progress [22]). However it would require a possible change of the Fredholm set-up of the current method and a good understanding of the Seiberg-Witten theory on any four manifold with a *degenerated symplectic form* as in Taubes program ([29], [30]). In the special case when Σ is disjoint from $\{n = 0\}$, Theorem 13 is obviously true as our analysis only involves the local geometry of Σ in M.

Third, if we do not restrict to instantons of small volume, then we have to take into account *bubbling* phenomena as in the pseudo-holomorphic curves case, and gluing of instantons of big and small volumes similar to [25] in Floer trajectory case. Nevertheless, one does not expect bubbling can occur when volumes of instantons are small, thus they would converge to a J_n-holomorphic curve in C as $\varepsilon \to 0$.

Last, we expect our result still holds in the *almost* G_2 setting, namely Ω is only a closed form rather than a parallel form. This is because our gluing analysis relies mainly on the Fredholm regularity property of the linearized instanton equation. Such a flexibility could be useful for finding regular holomorphic curves $\Sigma \subset C$ in order to apply our theorem.

References

[1] B. S. Acharya, B. Spence, *Supersymmetry and M theory on 7-manifolds,* [hep-th/0007213].

[2] M. Aganagic, C. Vafa, *Mirror Symmetry and a G_2 Flop,* J. High Energy Phys. 2003, no. 5, 061, 16 pp.

[3] S. Akbulut, S. Salur, *Deformations in G_2 manifolds,* Advances in Mathematics Volume 217, Issue 5, 20 March 2008, p2130–2140.

[4] C. Beasley, E. Witten, *A Note on Fluxes and Superpotentials in M-theory Compactifications on Manifolds of G_2 Holonomy,* Journal of High Energy Physics, Volume 2002.

[5] B. Booss-Bavnbek and K. Wojciechowski, *Elliptic Boundary Problems For Dirac Operators,* Birkhäuser (1993).

[6] R. Bryant, *Calibrated embeddings in the special Lagrangian and coassociative cases.* Special issue in memory of Alfred Gray (1939–1998). Ann. Global Anal. Geom. 18 (2000), no. 3-4, 405–435.

[7] A. Corti, M. Haskins, J. Nordstrom, T. Pacini, *G_2 manifolds and associative submanifolds via semi-Fano 3-folds,* arXiv: 1207.4470.

[8] K. Fukaya, Y.G. Oh, *Zero-loop open strings in the cotangent bundle and Morse homotopy.* Asian J. Math. 1 (1997), no. 1, 96–180.

[9] K. Fukaya, Y.G. Oh, H. Ohta, K. Ono, *Lagrangian intersection Floer theory - anomoly and obstruction,* AMS/IP Studies in Advanced Mathematics, International Press.

[10] D. Gayet, F. Witt, *Deformations of associative submanifolds with boundary,* Adv. Math. 226 (2011), 2351–2370.

[11] D. Gilbarg and N. Trudinger, *Elliptic Partial Differential Equations of Second Order,* 2nd Edition, Springer Verlag (1983).

[12] R. Gopakumar, C. Vafa, *M-theory and topological strings - II,* [hep-th/9812127].

[13] R. Harvey, B. Lawson, *Calibrated geometries,* Acta Math. 148 (1982), 47-157.

[14] H.A. Harvey, G. Moore, *Superpotentials and membrane instantons,* [hep-th/9907026].

[15] N. Hitchin, *The geometry of three forms in 6 and 7 dimensions, J. Differential Geom. 55 (2000), no. 3,* 547–576.

[16] N. Hitchin, *Stable forms and special metrics,* Comtemporary Math. 288 (2001), 70–89.

[17] D. D. Joyce, *Compact Riemannian 7-manifolds with holonomy G2. I, II,* J. Diff. Geom. 43 (1996), no. 2, 291-328, 329-375.

[18] D. D. Joyce, *Compact manifolds with special holonomy,* Oxford Mathematical Monographs, Oxford University Press, Oxford, 2000.

[19] A. Kovalev, *Twisted connected sums and special Riemannian holonomy*, J. Reine Angew. Math. 565 (2003), 125–160.
[20] J.H. Lee and N.C. Leung, *Instantons and Branes in Manifolds with Vector Cross Product* , Asian J. Math. 12, (2008), no. 1, 121–143.
[21] N.C. Leung, X. Wang, K. Zhu, *Thin instantons in G_2-manifolds and Seiberg-Witten invariants,* to appear in J. Diff. Geom., also available at arXiv:1107.1947 [math.DG].
[22] N.C. Leung, X. Wang, K. Zhu, *Thin instantons in G_2-manifolds with boundaries on intersecting coassociative submanifolds,* in preparation.
[23] D. McDuff, D. Salamon, *J-holomorphic Curves and Symplectic Topology,* Colloquim Publications, vol 52, AMS, Providence RI, 2004.
[24] R.C. McLean, *Deformations of Calibrated Submanifolds,* Commun. Analy. Geom. **6** (1998) 705–747.
[25] Y. Oh, K. Zhu, *Thick-thin decomposition of Floer trajectories and adiabatic gluing,* arXiv:1103.3525.
[26] C.Taubes, *Gr=>SW From pseudo-holomorphic curves to Seiberg-Witten solutions,* J. Differential Geom. 51 (1999), no. 2, 203–334.
[27] C.Taubes, S*W=>Gr: From the Seiberg-Witten equations to pseudo-holomorphic curves,* J. Amer. Math. Soc, Vol 9, no. 3, July 1996, 845–918.
[28] C.Taubes, *Gr=SW: Counting curves and connections,* J. Differential Geom. 52 (1999), no. 3, 453–609.
[29] C. Taubes, *The geometry of the Seiberg-Witten invariants.* Proceedings of the International Congress of Mathematicians, Vol. II (Berlin, 1998). Doc. Math. 1998, Extra Vol. II, 493–504.
[30] C. Taubes, *Seiberg-Witten invariants and pseudo-holomorphic subvarieties for self-dual, harmonic 2-forms.* Geom. Topol. 3 (1999), 167–210.
[31] E. Witten, *Supersymmetry and Morse theory,* J. Differential Geom. Volume 17, Number 4 (1982), 661–692.
[32] S.-T. Yau *On Ricci curvature of a compact Kähler manifold and complex Monge-Ampére equation I.* Comm. Pure and App. Math. 31, pp. 339–411, 1978.

The Institute of Mathematical Sciences, The Chinese University of Hong Kong, Shatin, Hong Kong
E-mail address: `leung@ims.cuhk.edu.hk`

Department of Mathematics, Rutgers University-Newark, Newark, NJ 07102
E-mail address: `xiaowwan@rutgers.edu`

Department of Mathematics, Harvard University, Cambridge, MA 02138
E-mail address: `kzhu@math.harvard.edu`

Proceedings of 19th Gökova
Geometry-Topology Conference
pp. 111 – 122

Locally conformally flat and self-dual structures on simple 4-manifolds

Mustafa Kalafat

Abstract. This is a survey article on the existence of locally conformally flat (LCF) and self-dual (SD) metrics on various basic 4-manifolds like simply-connected ones or product types.

1. Introduction

A Riemannian manifold (M, g) is called *locally conformally flat (LCF)* if there is a neighborhood U of any point, and a strictly positive smooth function f such that $\tilde{g} = fg$ is a metric of zero sectional curvature everywhere on $U \subset M$. Sometimes the more concise terminology *conformally flat* is also used since globally conformally flat manifolds admit flat metrics. They are quotients of $\mathbb{R}^n$ so that there is no need to give them a new name. Alternating and symmetry properties imply that we can think of the $(0, 4)$ Riemann curvature tensor R as an element of the space $S^2\Lambda^2 M \subset \otimes^4 T^*M$. It also satisfies the algebraic Bianchi identity, hence it lies in the kernel of the *Bianchi symmetrization map*

$$b : S^2\Lambda^2 M \to S^2\Lambda^2 M, \qquad b(T)(x, y, z, t) := \frac{1}{3} T((x, y, z), t).$$

Since $b^2 = b$ and b is $GL(T^*M)$-equivariant, we have the equivariant decomposition $S^2\Lambda^2 M = \operatorname{Ker} b \oplus \operatorname{Im} b$, where we call this kernel as the space of *curvature-like tensors*. Thinking $\operatorname{Ker} b$ as an invariant $O(g)$-module, we have a unique irreducible decomposition, according to this the curvature tensor decomposes as $R = U \oplus Z \oplus W$. The components can be computed as

$$U = \frac{s}{2n(n-1)} g \owedge g \quad \text{and} \quad Z = \frac{1}{n-2} \overset{\circ}{\mathrm{Ric}} \owedge g$$

where s is the scalar curvature, $\overset{\circ}{\mathrm{Ric}} = \mathrm{Ric} - \frac{s}{n} g$ is the trace-free Ricci tensor, "$\owedge$" is the *Kulkarni-Nomizu product* defined by,

$$A \owedge B\,(X, Y, Z, T) := \begin{vmatrix} A(X,Z) & B(X,T) \\ A(Y,Z) & B(Y,T) \end{vmatrix} + \begin{vmatrix} B(X,Z) & A(X,T) \\ B(Y,Z) & A(Y,T) \end{vmatrix},$$

Key words and phrases. Locally conformally flat manifolds, self-dual manifolds, Kähler manifolds.

which is commutative and multiplies two symmetric 2-tensors to produce a curvature-like 4-tensor. Finally W is the *Weyl tensor* which is defined to be what is left over from the first two pieces.

Next, assume that we are in dimension $n = 4$ and our manifold is oriented. Then the metric together with the orientation determines a unique volume form ω_g. In this case we define the *Hodge star involution* $*_g : \Lambda^2 \to \Lambda^2$ pointwise by imposing the equality $\langle\alpha, \beta\rangle\omega_g = \alpha \wedge *\beta$. This yields the ± 1 eigenspace decomposition $\Lambda^2 = \Lambda^2_+ \oplus \Lambda^2_-$ of the 2-forms. These 3-dimensional eigenspaces are interchanged if one works with the reversed orientation. By an appropriate change of indices, we consider $W : \Lambda^2 \to \Lambda^2$ as an operator. One can show that the mixed parts $W^-_+ : \Lambda^2_+ \to \Lambda^2_-$ and $W^+_- : \Lambda^2_- \to \Lambda^2_+$ vanish so that the Weyl tensor decomposes as $W = W^+_+ \oplus W^-_-$. Abbreviating $W_\pm = W^\pm_\pm$, we say that the Riemannian manifold is *self-dual* if $W_- \equiv 0$, *anti-self-dual* if $W_+ \equiv 0$ respectively. We also call each of these two cases as *half-conformally flat* if we do not want to specify any orientation. This terminology actually comes from the interpretation that W is a conformally invariant tensor if one considers it as a $(1, 3)$ tensor. One can show that the manifold is conformally flat if and only if the tensor $W \equiv 0$ for dimensions $n \geq 4$. See [Kü] for a proof. In dimension $n = 3$ local conformal flatness is determined by the Schouten tensor, and in dimension $n = 2$ all manifolds are locally conformally flat. Basic examples of LCF manifolds are constant sectional curvature spaces, e.g., S^n and T^n, their products with S^1 or $\mathbb{R}$, and products of two Riemannian manifolds with constant sectional curvature 1 and -1 respectively. See [Bes] for further details. Note that SD metrics are only defined on orientable 4-manifolds. Basic SD 4-manifolds are first of all LCF spaces, e.g. S^4, T^4, $S^1 \times S^3$ (see Theorem 3.10), among non-LCF ones we have $\mathbb{CP}_2$ with its Fubini-Study metric (see Corollary 3.1) and scalar-flat-Kähler (SFK) surfaces (see Theorem 2.1) in particular the K3 surface. SD 4-manifolds are first considered by [Pen] and later put on a firm mathematical foundation in [AHS].

In this survey, we analyze LCF and SD structures on various simple 4-manifolds like product type or simply-connected. These results are spread out various places. Some of them are not written, if so not in detail. We hope that it is a good public service to accumulate these results in an article. Interested reader may consult to the resources [AK] and [AKO] for some recent progress on this type of geometry. In section 2 we introduce the basic tools, in section 3 we apply these tools on the manifolds and finally in the appendix we present a partial result along with an open problem.

Acknowledgements. Thanks to S. Finashin, C. Koca and M. Korkmaz for useful discussions.

2. Tools

In this section we develop the main tools to analyze our spaces. Let M be a closed, oriented 4-manifold. We have the following two basic identities which connect quantities related to curvature with topological numbers.

$$\chi(M) = \frac{1}{8\pi^2}\int_M \frac{s^2}{24} - \frac{|\overset{\circ}{\mathrm{Ric}}|^2}{2} + |W_+|^2 + |W_-|^2\omega_g \tag{1}$$

$$\tau(M) = \frac{1}{12\pi^2}\int_M |W_+|^2 - |W_-|^2\omega_g \tag{2}$$

The first one is called the *generalized Gauss-Bonnet theorem* [AW, ST] which can also be generalized to all even dimensions. The second one is called the *signature formula* which is specific to dimension 4. It is obtained through the Hirzebruch signature theorem [Hi] $\tau(M) = \frac{1}{3}p_1[M]$ by expanding the Pontrjagin class with Chern-Weil theory [ST]. See also second volume of [KN]. These two equations relate geometric information with the topological ones. As an immediate application for example if a 4-manifold admits a locally conformally flat metric then both of the self-dual and anti-self-dual Weyl curvatures vanish since $W = W_+ \oplus W_- = 0$. Hence the signature formula (2) implies that the signature $\tau(M) = 0$. Perhaps this signature condition is the most important topological obstruction for admitting LCF metrics. We start with a very common way of producing ASD metrics on a Kähler manifold. Recall that a Riemannian manifold is called *scalar-flat* if its scalar curvature is zero everywhere.

Theorem 2.1. *A Kähler surface is scalar-flat(SF) iff anti-self-dual(ASD).*

Proof. We follow [Bes] and [LS]. The alternating and symmetry properties of the curvature tensor implies that it is a symmetric section of the bundle $\wedge^2 \otimes \wedge^2$. Furthermore, since we are on a Kähler manifold we have the identities [KN],

$$R(JX,JY)Z = R(X,Y)Z \quad \text{and} \quad R(X,Y)JZ = JR(X,Y)Z.$$

These imply respectively the J-invariance of the first and the second pair of components of the curvature tensor hence it is a type $(1,1)$ real 2-form in both of these components. So we can think of the Kähler curvature tensor as a symmetric section of $\wedge^{1,1} \otimes \wedge^{1,1}$ or after dualizing the second component, as a symmetric element of $\mathrm{End}(\wedge^{1,1})$.

Since we are on a complex manifold, we have the Dolbeault decomposition of 2-forms $\wedge^2_{\mathbb{C}} = \wedge^{2,0} \oplus \wedge^{1,1} \oplus \wedge^{0,2}$ according to their type. This decomposition is orthogonal with respect to the action of the Hodge star operator $*_g$. We also have the orthogonal eigenspace decomposition in dimension 4 as explained in the introduction. In the Kähler case, if one complexifies these eigenspaces, we claim to have the following:

$$\wedge^2_{+\mathbb{C}} = \mathbb{C}\omega \oplus \wedge^{2,0} \oplus \wedge^{0,2}$$

$$\wedge^2_{-\mathbb{C}} = \wedge^{1,1}_0$$

Here, the set of *primitive* $(1,1)$ *forms* $\wedge^{1,1}_0$ can be defined to be the orhogonal complement of ω in $\wedge^{1,1}$. Take a unitary coframe $\{dz^1, dz^2\}$ at a point. Then the Kähler form is

$$\omega = \frac{i}{2}\sum_{i=1}^{2} dz^i \wedge d\bar{z}^i = dx^1 \wedge dy^1 + dx^2 \wedge dy^2$$

and the volume form is computed as

$$\omega_g = \frac{\omega^2}{2!} = \frac{-1}{4} dz^{1\bar{1}2\bar{2}} = dx^1 \wedge dy^1 \wedge dx^2 \wedge dy^2.$$

Since $*_g(dx^i \wedge dy^i) = dx^{3-i} \wedge dy^{3-i}$ we have $*_g\omega = \omega$, i.e., the Kähler form is self-dual. This observation links to the following interpretation of the primitive $(1,1)$ forms. We can write $\Lambda_0^{1,1} = \operatorname{Ker} L$, for the *Lefschetz operator* $L : \Lambda^{1,1} \to \Lambda^{2,2}$ defined by $L(\alpha) = \alpha \wedge \omega$ since

$$\langle \alpha, \omega \rangle = 0 \Leftrightarrow 0 = \langle \alpha, \omega \rangle \omega_g = \alpha \wedge *_g\omega = \alpha \wedge \omega = L(\alpha).$$

The forms $\omega_5 = dz^1 \wedge dz^2$ and $\omega_6 = d\bar{z}^1 \wedge d\bar{z}^2$ pointwise generate the complex bundles $\Lambda^{2,0}$ and $\Lambda^{0,2}$ respectively. One easily checks that their real parts are self-dual. Hence ω, ω_5, $\omega_6 \in \Lambda^2_{+\mathbb{C}}$ are orthogonal and their real parts are elements of the real rank 3 vector bundle of self-dual 2-forms Λ^2_+. On the other hand the $(1,1)$ forms $\omega_2 = dz^1 \wedge d\bar{z}^2$, $\omega_3 = dz^2 \wedge d\bar{z}^1$ and $\omega_4 = dz^1 \wedge d\bar{z}^1 - dz^2 \wedge d\bar{z}^2$ are orthogonal and their real parts are anti-self-dual. For example $*_g\Re\omega_2 = *_g(dx^{12} + dy^{12}) = -(dx^{12} + dy^{12}) = -\Re\omega_2$. Moreover they are all orthogonal to ω, ω_5, $\omega_6 \in \Lambda^2_{+\mathbb{C}}$.

Now, since the curvature operator $\mathcal{R}$ is in $\operatorname{End}(\Lambda^{1,1})$, its upper left piece $\mathcal{R}^+_+ = \mathcal{W}_+ + \frac{s}{12}I$ is an element of $\operatorname{End}(\mathbb{C}\omega)$. So suppose $\mathcal{R}^+_+ = f\,\omega \otimes \omega^\sharp$ for some function $f : M \to \mathbb{R}$.

$\mathcal{R}$	$\mathbb{C}\omega \oplus \Lambda^{2,0} \oplus \Lambda^{0,2}$	$\Lambda_0^{1,1}$
$\mathbb{C}\omega \oplus \Lambda^{2,0} \oplus \Lambda^{0,2}$	$\mathcal{W}_+ + \frac{s}{12}I$	$\overset{\circ}{\text{Ric}}$
$\Lambda_0^{1,1}$	$\overset{\circ}{\text{Ric}}$	$\mathcal{W}_- + \frac{s}{12}I$

TABLE 1. Curvature operator for Kähler surfaces

$\mathcal{R}$	$\mathbb{C}\omega$	$\Lambda_0^{1,1}$
$\mathbb{C}\omega$	$s/4\cdot$	$\overset{\circ}{\text{Ric}}$
$\Lambda_0^{1,1}$	$\overset{\circ}{\text{Ric}}$	$\mathcal{W}_- + \frac{s}{12}I$

$\mathcal{R}^+_+$	$\mathbb{C}\omega$	$\Lambda^{2,0}$	$\Lambda^{0,2}$
$\mathbb{C}\omega$	$s/4\cdot$	0	0
$\Lambda^{2,0}$	0	0	0
$\Lambda^{0,2}$	0	0	0

TABLE 2. Curvature operator and its self-dual part for Kähler surfaces

To figure out the function f we need to compute some inner products. First compute the norm of the Kähler form ω. Since the volume form $\omega_g = \omega^2/2!$ we have

$$\langle \omega, \omega \rangle\, \omega_g = \omega \wedge *\omega = \omega \wedge \omega = 2\,\omega_g.$$

Hence we get $|\omega| = \sqrt{2}$. [1] Secondly we want to compute the inner product $\langle \rho, \omega \rangle$ where ρ is the *Ricci form* defined by $\rho(\cdot,\cdot) = \mathrm{Ric}(J\cdot,\cdot)$. The above trick does not work in this case. We need to use some tensor analysis. Start with fixing a convention for the complex structure. Suppose $J = J_i^{\ k}\, dx^i \otimes \partial_k$. Then the basic property $J^2 = -Id$ reads $J_i^{\ k} J_k^{\ j} = -\delta_i^j$ [2] in terms of tensors. Keep in mind that the ω and J are skew symmetric tensors. We compute the following:

$$\omega^{ij} = \omega(dx^i, dx^j) = g(J dx^i, dx^j) = g(J_k^{\ i} dx^k, dx^j) = J_k^{\ i} g^{kj} = J^{ji} = -J^{ij}.$$

$$\rho_{ij} = \rho(\partial_i, \partial_j) = \mathrm{Ric}(J\partial_i, \partial_j) = J_i^{\ k}\mathrm{Ric}(\partial_k, \partial_j) = J_i^{\ k} R_{kj}.$$

And then,

$$\begin{aligned} \langle \rho, \omega \rangle &= \frac{1}{2!}\rho_{ij}\,\omega^{ij} = \frac{1}{2} J_i^{\ k} R_{kj}(-J^{ij}) = \frac{-1}{2} J_i^{\ k} R_{kj} J_k^{\ j} g^{ki} \\ &= \frac{-1}{2}(-\delta_i^j) R_{kj} g^{ki} = \frac{1}{2} R_{ki} g^{ki} = \frac{s}{2}. \end{aligned}$$

Now, writing $\mathcal{R}\omega = f\omega + g\omega^{\perp}$, multiplying both sides with the Kähler form and using $\mathcal{R}\omega = \rho$, we get the following:

$$\begin{aligned} \langle \mathcal{R}\omega, \omega \rangle &= f\,\langle \omega, \omega \rangle \\ \langle \rho, \omega \rangle &= f\,\langle \omega, \omega \rangle \\ s/2 &= f \cdot 2 \\ s/4 &= f. \end{aligned}$$

Hence $\mathcal{W}_+$ is a multiple of the scalar curvature s. Therefore we have $s = 0$ if and only if $\mathcal{W}_+ = 0$. □

This result has an immediate corollary.

Corollary 2.2. *For Kähler metrics on a complex surface we have the pointwise identity*

$$|W_+|^2 = \frac{s^2}{24}. \tag{3}$$

Proof. Since in the Kähler case the term $\mathcal{R}_+^+ = W_+ + \frac{s}{12} I$ acts by the following

$$\frac{s}{4}\begin{bmatrix} 1 & 0 & 0 \\ 0 & 0 & 0 \\ 0 & 0 & 0 \end{bmatrix},$$

as in the Table 2, we have

$$W_+ = \frac{s}{12}\begin{bmatrix} 2 & 0 & 0 \\ 0 & -1 & 0 \\ 0 & 0 & -1 \end{bmatrix}.$$

[1] In general dimension n, writing in local orthonormal frame $\omega = e_1 \wedge e_2 + \ldots + e_{2n-1} \wedge e_{2n}$ we compute $|\omega| = \sqrt{n}$. Then from $|\omega|^2 \omega_g = \omega \wedge *\omega$ we get $*\omega = |\omega|^2 \omega^{n-1}/n!$ so $*\omega = \omega^{n-1}/(n-1)!$

[2] If one fixes the alternative convention $J = J_i^k\, \partial_k \otimes dx^i$ one gets $J_{\ k}^{j} J_{\ i}^{k} = -\delta_i^j$ and $\omega_{ij} = -J_{ij}$, a negative sign.

Taking the norms of both sides yields the result. □

Theorem 2.3. *If a 4-manifold M admits a Kähler-Einstein (KE) metric g then $\chi(M) = 3\tau(M)$ if and only if g is self-dual (SD).*

Proof. For Kähler metrics we have the pointwise identity (3)

$$|W_+|^2 = \frac{s^2}{24}.$$

For Einstein metrics, by definition the trace-free Ricci tensor $\overset{\circ}{\mathrm{Ric}} = 0$. Plugging these two identities into the Gauss-Bonnet formula (1) we get

$$8\pi^2\chi = 2\,\|W_+\|^2 + \|W_-\|^2. \tag{4}$$

Eliminating $\|W_+\|$ from (2) and (4) we get the following equality

$$8\pi^2(\chi - 3\tau) = 3\,\|W_-\|,$$

which yields the result. □

Finally we state the celebrated theorems of Kuiper. See also [Ho] for a recent exposition and improvement.

Theorem 2.4 ([Kui]). *Let (M^n, g) be a simply connected, LCF n-manifold of class C^1. Then there is a conformal immersion $f : M \to S^n$. If in addition M is compact, then this map is a conformal diffeomorphism.*

Here comes another very useful theorem of Kuiper. See also [Kob].

Theorem 2.5 ([Kui2]). *Universal cover of a compact, LCF space with an infinite Abelian fundamental group must be $\mathbb{R}^n$ or $\mathbb{R} \times S^{n-1}$.*

3. Simple 4-manifolds

We start with a basic space, the complex projective space with its standart Fubini-Study metric. This can be though as the metric quotient of $S^{2n+1} \subset \mathbb{C}^{n+1}$ by unit complex scalar multiplication. Alternatively on $\mathbb{C}^n$ take the following complex metric coefficients:

$$g_{FS_{i\bar{j}}} = g_{FS}(\partial_i, \partial_{\bar{j}}) := \frac{(1+|z|^2)\delta_{i\bar{j}} - z_i z_{\bar{j}}}{(1+|z|^2)^2}.$$

Taking the completion of this space gives the complex projective space $\mathbb{CP}_n$. As an application of the theorems in the previous section we obtain the following.

Corollary 3.1. $(\mathbb{CP}_2, g_{FS})$, *the complex projective space with its Fubini-Study metric is self-dual.*

Proof. One can easily compute $\chi(\mathbb{CP}_2) = 3$. Since the intersection form $Q_{\mathbb{CP}_2} = [1]$ we have $\tau = 1$. These satisfy the equality in Theorem 2.3. □

Theorem 3.2. *The underlying smooth manifold of* $\mathbb{CP}_2$*, the complex projective space does not admit any LCF metrics.*

Proof. The basic obstruction signature $\tau(\mathbb{CP}_2) = 1$ is nontrivial. □

Next we work on the 4-manifold $S^2 \times S^2$, the product of two spheres. The spheres $S^2 \times q$ and $p \times S^2$ generating the homology have self-intersection zero, and $+1$ with each other. So

$$Q_{S^2 \times S^2} = \begin{bmatrix} 0 & 1 \\ 1 & 0 \end{bmatrix},$$

called the hyperbolic matrix and denoted by H. It has eigenvalues ± 1 and hence the signature $\tau = 0$. [3] Hence an obstruction vanishes for locally conformally flatness. However, this turns out to be not sufficient as follows.

Theorem 3.3. *The 4-manifold* $S^2 \times S^2$ *does not admit any LCF nor even SD metrics.*

Proof. Since by Kuiper's theorem [Kui], any compact, simply-connected, LCF Riemannian 4-manifold is conformally equivalent to the round 4-sphere, $S^2 \times S^2$ does not admit any LCF metric. Suppose it does have a self-dual metric. Then $W_- = 0$ and by the signature formula (2) the integral $0 = \int_M |W_+|^2 \omega_g$. So, pointwise $W_- = 0$ and $W = 0$. This yields a LCF metric which is already a contradiction by the previous argument. □

This example also illustrates the fact that the product of LCF manifolds may not be LCF.

Corollary 3.4. *The 4-manifold* $K3$ *is not LCF, but SD.*

Proof. Since the intersection form of a K3 surface is $Q = 2E_8 \oplus 3H$, the signature $\tau(K3)$ is -16 which is nonzero so that it can not be LCF. Since the first Chern class $2\pi c_1(K3)$ is zero, the zero form is a $(1,1)$-form representing this class. Since $K3$ is compact and Kähler by Yau's resolution to the Calabi's problem [Yau], there is a Kähler metric g in the same class with Ricci form $\rho_g \equiv 0$. Hence g is Ricci flat, so scalar flat. SFK surfaces are ASD, hence SD with the reversed orientation. □

Theorem 3.5. *The smooth* 4*-manifold* $\mathbb{CP}_2 \# \overline{\mathbb{CP}}_2$ *does not admit any LCF nor SD metrics.*

Proof. Since $\tau = 0$, the argument in the proof of Theorem 3.3 is again valid for this 4-manifold. □

A similar argument also excludes the manifolds $k\mathbb{CP}_2 \# k\overline{\mathbb{CP}}_2$ for any $k > 0$. Since we used it multiple times, it is convenient to sum up the idea of the proof as follows.

[3] Alternatively, the map $(I_3, -I_3)$ is an orientation reversing diffeomorphism of $S^2 \times S^2$. So Hirzebruch signatures mapped onto each other and $\tau = -\tau$.

Theorem 3.6. *Let M be a compact, oriented 4-manifold with signature $\tau = 0$, then we have the following.*

1. A metric g on M is SD iff LCF.

2. If M is simply connected but not diffeomorphic to S^4 then it does not admit any LCF or SD metric.

Theorem 3.7. *The smooth 4-manifold $S^1 \times S^1 \times S^1 \times S^1 = T^2 \times T^2$ is LCF hence SD.*

Proof. Since T^4 is a quotient of $\mathbb{R}^2$ by $\mathbb{Z}^2$ isometries, it is flat hence LCF and consequently SD.

The second assertion is alternatively seen as follows: The product of flat metrics on the components is scalar-flat-Kähler (SFK). So by the Theorem 2.1 it is ASD. □

As a by-product of local conformal flatness we also compute $\tau(T^4) = 0$. Bieberbach's theorem states that the only compact manifolds that admit flat metrics are T^n and its finite quotients. Hence these are the only (globally) conformally flat manifolds. Since these are already classified, the phrase "conformally flat" is usually used in place of LCF in the literature. Before checking another possible basic product in dimension four, let us look at its universal cover.

Theorem 3.8. *The smooth 4-manifold $S^2 \times \mathbb{R}^2$ is LCF hence SD.*

Proof. Since $S^2 \times \mathbb{R}$ is diffeomorphic to $\mathbb{R}^3 - \{0\}$, the inherited standart metric is of constant (positive) sectional curvature. A constant sectional curvature space times $\mathbb{R}$ is LCF. □

Next result indicates that the above manifold can not have isometries that give $S^2 \times T^2$ as the quotient.

Theorem 3.9. *The smooth 4-manifold $S^2 \times T^2$ is neither LCF nor SD.*

Proof. Because of the antipodal orientation reversing isometry of the 2-sphere, we have the signature $\tau(S^2 \times T^2) = 0$. Hence this is LCF iff SD. Secondly, since this manifold has Abelian infinite fundamental group $\mathbb{Z}^2$, if it admits a LCF metric then applying Kuiper's second theorem (Theorem 2.5) it should have universal cover $\mathbb{R}^4$ or $\mathbb{R} \times S^3$. Since the universal cover is $S^2 \times \mathbb{R}^2$ this gives a contradiction. In general this argument applies to $S^p \times T^q$ for $p, q \geq 2$. □

This manifold is particularly interesting since even though it is not LCF, the following infimum of all the possible Weyl energies on the manifold,

$$W(M) := \inf\left\{ \int_M |W_g|_g^2 \omega_g : g \in \mathcal{M}_M \right\}$$

called the *Weyl invariant* is zero, where W_g is the conformally invariant type $(1,3)$ Weyl tensor, $\mathcal{M}_M$ is the space of smooth metrics. See [Kob] for a proof. Hence the infimum is not attained.

One can easily compute the Euler characteristic of $S^1 \times S^3$ to be $\chi = 0$, since the S^3 component is an oriented closed 3-manifold hence $\chi(S^3) = 0$. Since there is no second homology, the signature $\tau = 0$. In this case it is a candidate of a manifold which may admit LCF metrics. This turns out to be the case as follows.

Theorem 3.10. *The 4-manifold $S^1 \times S^3$ with its standard product metric is LCF and hence SD.*

Proof. The proof is adapted from [JV]. Locally we can think this as $\mathbb{R} \times S^3$, where $\mathbb{R}$ is the flat line with $g_{\mathbb{R}} = dx \otimes dx$ and S^3 has the round metric of curvature $+1$. Rm denoting the Riemann curvature tensor, $\owedge$ is the Kulkarni-Nomizu product which is commutative,

$$\begin{aligned}
\mathrm{Rm}_{\mathbb{R}\times S^3} &= \mathrm{Rm}_{\mathbb{R}} + \mathrm{Rm}_{S^3} \\
&= 0 + \tfrac{1}{2}\, g_{S^3} \owedge g_{S^3} \\
&= \tfrac{1}{2}\,(g_{S^3} + dx^2) \owedge (g_{S^3} - dx^2) \\
&= \tfrac{1}{2}\, g_{\mathbb{R}\times S^3} \owedge (g_{S^3} - dx^2) \\
&= \Psi(\tfrac{1}{2}\,(g_{S^3} - dx^2))
\end{aligned}$$

where by the definition of the product $dx^2 \owedge dx^2 = 0$, and $\Psi : S^2(T^*M) \longrightarrow \operatorname{Ker} b$ is defined by $\Psi(h) = h \owedge g_M$. Here, $\operatorname{Ker} b$ is the space of curvature-like tensors as in the introduction. The decomposition $\operatorname{Ker} b = \mathcal{W} \oplus \Psi(S_0^2(T^*M)) \oplus \Psi(\mathbb{R}g)$ implies that the Weyl tensor vanishes. □

Theorem 3.11. *The 4-manifolds $S^2 \times \Sigma_g$ with their standard product metric is LCF and hence SD for $g \geq 2$.*

Proof. We adapt from [Bes]. Suppose we have the constant sectional curvature $+1$ metric on S^2 and -1 metric on Σ_g. Then we have the following descriptions

$$\mathrm{Rm}_{S^2} = \frac{1}{2}\, g_{S^2} \owedge g_{S^2} \quad \text{and} \quad \mathrm{Rm}_{\Sigma_g} = \frac{-1}{2}\, g_{\Sigma_g} \owedge g_{\Sigma_g}$$

for the Riemann curvature tensors.

$$\begin{aligned}
\mathrm{Rm}_{S^2\times\Sigma_g} &= \mathrm{Rm}_{S^2} + \mathrm{Rm}_{\Sigma_g} \\
&= \tfrac{1}{2}\left(g_{S^2} \owedge g_{S^2} - g_{\Sigma_g} \owedge g_{\Sigma_g}\right) \\
&= \tfrac{1}{2}\left(g_{S^2} + g_{\Sigma_g}\right) \owedge \left(g_{S^2} - g_{\Sigma_g}\right) \\
&= \tfrac{1}{2}\, g_{S^2\times\Sigma_g} \owedge \left(g_{S^2} - g_{\Sigma_g}\right) \\
&= \Psi\big(\tfrac{1}{2}\left(g_{S^2} - g_{\Sigma_g}\right)\big).
\end{aligned}$$

Being in the image of Ψ, the Weyl tensor vanishes. Alternatively, starting with the Kähler metrics, one obtains a scalar-flat-Kähler (SFK) metric on the product. Now apply Theorem 2.1 with both orientations. □

Appendix A.

Using Gauss-Bonnet, and signature formula techniques we can also prove the following.

Theorem A.1. *If a 4-manifold admits a Kähler-Einstein(KE) metric which is also locally conformally flat(LCF) then its Euler characteristic $\chi = 0$.*

Proof. Recall the pointwise identity of Corollary 2.2 for Kähler metrics

$$|W_+|^2 = \frac{s^2}{24}.$$

For Einstein metrics, by definition the trace-free Ricci tensor $\overset{\circ}{\mathrm{Ric}} = 0$. Plugging these two identities into the Gauss-Bonnet formula (1) we get

$$8\pi^2\chi(M) = 2\,\|W_+\|^2 + \|W_-\|^2.$$

Locally conformally flatness implies $W_\pm = 0$, hence $\chi = 0$ by above. □

As an application we can prove the following non-existence result.

Theorem A.2. *The product metric on the 4-manifolds $\Sigma_g \times \Sigma_h$, product of surfaces of genus $g, h \geq 2$ is not a LCF nor SD metric.*

Proof. First of all, $\Sigma_g \times \Sigma_h$ admits a Kähler-Einstein metric. One can see this through different ways. One is to use Aubin/Yau theorem, since $c_1 < 0$, a surface of general type, there exists a unique KE metric on this complex surface. Another way is to think in terms of product metrics. If you have the hyperbolic -1 curvature Kähler metrics on both components, the product metric is Kählerian. Besides that, the product of Einstein metrics with common cosmological constant is again Einstein. Combining the two, we obtain a Kähler-Einstein metric on the manifold. If the manifold admits a LCF metric as well, then Theorem A.1 implies that $\chi = 0$, however $\chi = (2-2g)(2-2h)$ which is a contradiction. So the product of two hyperbolic metrics on $\Sigma_g \times \Sigma_h$ is not LCF.

To analyze SD structure we need to find the signature of the manifold. A two dimensional oriented surfaces of genus g always has an orientation reversing diffeomorphism (involution) R_g. One can construct this by using a mirror reflection or reflection through a point after arranging the holes symmetrically. Then (R_g, I) is going to be an orientation reversing diffeomorphism of the 4-manifold. Since the Hirzebruch signature is diffeomorphism invariant and changing the orientation changes its sign, we have $\tau = -\tau$ hence $\tau(\Sigma_g \times \Sigma_h) = 0$. Alternatively one can compute the intersection matrix as $gh(-H_4) \oplus H_2$ and hence the characteristic polynomial $(\lambda^2-1)^{2gh}(\lambda^2-1)$. (Another approach might be exploiting only the parity of the intersection form and use Rokhlin's theorem of divisibility of the signature by 8 in the case of even intersection forms to obtain at least some of the cases.) If the product metric is a SD metric on the manifold then it is LCF by Theorem 3.6, which is already violated. □

Existence of LCF metrics on the product $\Sigma_g \times \Sigma_h$ of surfaces of genus $g \geq 2$ and $h \geq 1$ still remains as an open problem.

References

[AK] S. Akbulut, M. Kalafat, *A class of locally conformally flat 4-manifolds*, New York J. Math. 18 (2012) 733-763.

[AKO] H. Argüz, M. Kalafat, Y. Ozan, *Self-Dual metrics on non-simply connected 4-manifolds*, J. Geom. Phys. 64 (2013) 79-82.

[AW] Carl B. Allendoerfer, André Weil, *The Gauss-Bonnet theorem for Riemannian polyhedra*, Trans. Amer. Math. Soc. 53, 101-129. (1943).

[AHS] M. F. Atiyah, N. J. Hitchin, and I. M. Singer, *Self-duality in four-dimensional Riemannian geometry*, Proc. Roy. Soc. London Ser. A, 362 (1978), pp. 425–461.

[Bes] A. Besse, *Einstein Manifolds*, Springer-Verlag, 1987.

[Hi] F. Hirzebruch, *Topological methods in algebraic geometry*, Third enlarged edition, Springer-Verlag, New York, 1966 x+232 pp.

[Ho] Ralph Howard, *Kuiper's Theorem On Conformally Flat Manifolds.* Lecture Notes. Available at http://www.math.sc.edu/~howard/.

[Kob] Osamu Kobayashi, *On a conformally invariant functional of the space of Riemannian metrics.* J. Math. Soc. Japan 37 (1985), no. 3, 373-389.

[KN] S. Kobayashi and K. Nomizu, *Foundations of differential geometry 1-2*, Wiley & Sons, New York 1963, 1969.

[Kui] N. H. Kuiper, *On conformally-flat spaces in the large*, Ann. of Math. (2) 50, (1949). 916-924.

[Kui2] N. H. Kuiper, *On compact conformally Euclidean spaces of dimension* > 2, Ann. of Math. (2) 52, (1950). 478-490.

[Kü] Wolfgang Kühnel, *Differential geometry: Curves-surfaces-manifolds,* Second Edition, Student Mathematical Library, 16, AMS, Providence, RI, 2006.

[LeSD] C. LeBrun, *On the topology of self-dual 4-manifolds*, Proc. Amer. Math. Soc. 98 (1986), no. 4, 637–640.

[LS] C. LeBrun, M. Singer, *Existence and deformation theory for scalar-flat Kähler metrics on compact complex surfaces,* Invent. Math. 112 (1993), no. 2, 273313.

[Pen] R. Penrose, *Nonlinear gravitons and curved twistor theory*, The riddle of gravitation–on the occasion of the 60th birthday of Peter G. Bergmann (Proc. Conf., Syracuse Univ., Syracuse, NY., 1975), General Relativity and Gravitation 7 (1976), no. 1, 31–52.

[ST] I. M. Singer, J. A. Thorpe, *The curvature of 4-dimensional Einstein spaces*, Global Analysis (Papers in Honor of K. Kodaira) pp. 355–365, Univ. Tokyo Press, 1969.

[JV] J. Viaclovsky, *Lecture Notes on Differential Geometry*, available at author's website http://www.math.wisc.edu/~jeffv/, 2007.

[Yau] S.T. Yau, *On the Ricci curvature of a compact Kähler manifold and the complex Monge-Ampére equation I*, Comm. Pure Appl. Math. 31 (1978), no. 3, 339–411.

Department of Mathematics, Michigan State University, East Lansing, 48824, MI, USA
E-mail address: kalafat@math.msu.edu

Proceedings of 19th Gökova
Geometry-Topology Conference
pp. 123 – 145

Tube formula for self-similar fractals with non-Steiner-like generators

Ali Deniz, Şahin Koçak, Yunus Özdemir and Adem Ersin Üreyen

ABSTRACT. We give a pointwise proof for the tube formula of Lapidus-Pearse for self-similar fractals, where we extend the types of generators beyond the piecewise polynomial case. We illustrate the new approach on several examples.

1. Introduction

M. Lapidus and E. Pearse proved in [5] a tube formula for higher dimensional fractals, extending the earlier work of Lapidus and his coworkers [4]. They associate certain tilings with fractals and express the volume of the inner ε-neighborhood of the tiling in terms of residues of a certain associated function, called the tubular ζ-function. The residues thereby are taken at the complex dimensions of the fractal, a notion introduced and elaborated by Lapidus and his coworkers.

By a "tube formula" it is understood a formula giving the sum of Lebesgue measures of the inner ε-neighborhoods of the open sets constituting a tiling associated with a fractal as explained in Section 2. (The union of the inner ε-neighborhoods is also regarded as "the inner ε-neighborhood of the tiling".) It should be noted that this Lebesgue measure of the inner ε-neighborhood of the tiling is generally different from the Lebesgue measure of the ε-neighborhood of the fractal itself. But a recent work of E. Pearse and S. Winter [8] clarifies the relationship between the inner ε-neighborhood of the tiling and the genuine ε-neighborhood of the fractal in a very satisfactory way: If the boundary of the convex hull of the fractal is a subset of the fractal, then the volume of the ε-neighborhood of the fractal is the sum of the volumes of the inner ε-neighborhood of the tiling and the outer ε-neighborhood of the convex hull of the fractal.

As the volume of the outer ε-neighborhood of the convex hull is rather trivial, the tube formula in terms of residues gives effectively the true volume of the ε-neighborhood of the fractal if the Pearse-Winter condition is fulfilled. This circumstance attributes a higher value to the utility of the tube formula.

The original proof of the tube formula of Lapidus-Pearse was distributional, but thereafter pointwise proofs have been given in [2],[3] and [6].

Key words and phrases. Self-similar fractals, tube formula, complex dimensions, self-similar tiling, zeta functions.

In [2], we generalized part of their theory to the graph-directed fractals, defined complex dimensions for them and gave a general scheme of summation yielding tube formulas for self-similar as well as graph-directed fractals.

The original setting of Lapidus-Pearse assumed the generators of the tiling to be pluriphase, i.e., the volumes of the inner tubes of the generators should be piecewise polynomial. In [3] and [6] the theory has been extended to more general types of generators, but the conditions and the terminology differ in the two approaches. We consider generators of the following type: Volumes of the inner tubes of generators should be polynomial on a range between zero and a certain value less than the inradius and then arbitrarily piecewise continuously differentiable on the range between this intermediary value and the inradius. Historically, the adjective "Steiner" refers to the theorem that the volume of the ε-neighborhood of a convex body in $\mathbb{R}^d$ is a polynomial of ε. We prefer to use the term "Steiner-like" for a generator with a piecewise polynomial inner tube volume, as a synonym of "pluriphase" and reserve the term "non-Steiner-like" for other types of generators. We note that in [6] the term "Steiner-like" is used essentially for all (bounded) open sets. Our goal is to give a pointwise proof of a tube formula for self-similar fractals with non-Steiner-like generators of the above type and illustrate it on several examples.

2. The Tube Formula for Self-Similar Tilings

Let

$$F = \bigcup_{j=1}^{J} \varphi_j(F) =: \Phi(F) \subset \mathbb{R}^d$$

be a self-similar fractal, where $\varphi_j : \mathbb{R}^d \to \mathbb{R}^d$ are similitudes with scaling ratios $0 < r_j < 1$, $j = 1, \ldots, J$. Let $C := [F]$ be the convex hull of the fractal (for which we assume $\dim C = d$). We assume that the system $\{\varphi_j\}$ satisfies the so-called "tileset condition" of [8]: The open set condition with $\operatorname{int} C$ a feasible open set. Additionally we assume the non-triviality condition of [8]:

$$\operatorname{int} C \not\subseteq \Phi(C) \doteq \bigcup_{j=1}^{J} \varphi_j(C).$$

This condition amounts to $\operatorname{int} F = \emptyset$, [8, Proposition 2.11].

Now define $T_1 = \operatorname{int}(C \setminus \Phi(C))$ and its iterates $T_n = \Phi^{n-1}(T_1)$, $n = 2, 3, \ldots$ (see [7]). The tiling of the self-similar system is given by

$$\mathcal{T} := \{T_n\}_{n=1}^{\infty}$$

and the volume of the inner ε-neighborhood of the tiling $\mathcal{T}$ is defined by

$$V_{\mathcal{T}}(\varepsilon) := \sum_{n=1}^{\infty} V_{T_n}(\varepsilon),$$

where $V_{T_n}(\varepsilon)$ is the volume of the inner ε-neighborhood of T_n.

To state the tube formula we need some additional assumptions and definitions. Assume that T_1 is the union of finitely many (connected) components, $T_1 = G_1 \cup \cdots \cup G_Q$, called the generators of the tiling. In the original work of Lapidus-Pearse (L-P) they assume the generators to be Steiner-like in the following sense:

A bounded, open set $G \subset \mathbb{R}^d$ is called (monophase) Steiner-like if the volume $V_G(\varepsilon)$ of the inner ε-neighborhood of G admits an expression of the form

$$V_G(\varepsilon) = \sum_{i=0}^{d-1} \kappa_i(G)\varepsilon^{d-i}, \qquad \text{for } \varepsilon \leq g,$$

where g denotes the inradius of G, i.e., supremum of the radii of the balls contained in G.

For $\varepsilon > g$ we have $V_G(\varepsilon) = \text{volume}(G)$ which is denoted by $-\kappa_d(G)$, the negative sign being conventional [5].

To be precise, L-P consider also the "pluriphase" case, where $V_G(\varepsilon)$ is given piecewise by different polynomials in the region $0 < \varepsilon < g$. But this generalization brings no essential complication.

Lapidus-Pearse introduce the following "scaling ζ-function":

Definition 2.1. The scaling ζ-function of the self-similar fractal is defined by

$$\zeta(s) = \sum_{k=0}^{\infty} \sum_{w \in W_k} r_w^s,$$

where W_k is the set of words $w = w_1 w_2 \cdots w_k$ of length k (with letters from $\{1, 2, \ldots, J\}$) and $r_w = r_{w_1} r_{w_2} \ldots r_{w_k}$.

The above series can be shown to converge for $\text{Re}(s) > D$, where D is the *similarity dimension* of the system (i.e., the unique real root of the Moran equation $1 - \sum_{j=1}^{J} r_j^s = 0$ which coincides with the Minkowski and Hausdorff dimensions if the open-set condition holds). A simple calculation shows that $\zeta(s)$ can be expressed as [4, Theorem 2.4]

$$\zeta(s) = \frac{1}{1 - \sum_{j=1}^{J} r_j^s} \qquad \text{for } \text{Re}(s) > D. \tag{1}$$

$\zeta(s)$ can then be meromorphically extended to the whole complex plane. We will denote this extension also by $\zeta(s)$.

Definition 2.2. The set $\mathfrak{D} := \{\omega \in \mathbb{C} \mid \zeta(s) \text{ has a pole at } \omega\}$ is called the set of complex dimensions of the self-similar fractal.

Lapidus-Pearse define a second type of "ζ-function" associated with the tiling and related to the geometry of the (monophase) Steiner-like generators. As the case of multiple generators does not bring additional complications, we express their notion of tubular ζ-functions and their tube formula for the case of a single generator G.

Definition 2.3. The tubular ζ-function $\zeta_{\mathcal{T}}(s,\varepsilon)$ associated with the generator G is defined by

$$\zeta_{\mathcal{T}}(s,\varepsilon) := \zeta(s)\varepsilon^{d-s}\sum_{i=0}^{d}\frac{g^{s-i}}{s-i}\kappa_i(G).$$

The formula of Lapidus-Pearse for $V_{\mathcal{T}}(\varepsilon)$ now reads as follows:

Theorem 2.1 (Tube formula for tilings of self-similar fractals, [5]).

$$V_{\mathcal{T}}(\varepsilon) = \sum_{\omega\in\mathfrak{D}_{\mathcal{T}}} \operatorname{res}(\zeta_{\mathcal{T}}(s,\varepsilon);\omega), \tag{2}$$

where $\mathfrak{D}_{\mathcal{T}} = \mathfrak{D}\cup\{0,1,\ldots,d-1\}$.

Remark 2.1. A distributional proof for this formula is given in [5] that also holds for fractal sprays which is a more general concept than self-similar fractals. It is shown in [2] that the above formula holds pointwise for self-similar as well as graph-directed fractals. Pointwise proofs were also given in [3] and [6].

The aim of this work is to extend the Lapidus-Pearse tube formula to the case where the generators are non-Steiner-like in the sense below. The paper [6] deals with the same question where the authors use other conditions on the generators and a different terminology.

We shall assume the volume $V_G(\varepsilon)$ of the inner ε-neighborhood of a generator G to be a continuous function of the following form:

$$V_G(\varepsilon) = \begin{cases} \sum_{i=0}^{d-1}\kappa_i(G)\varepsilon^{d-i} & \text{for } 0<\varepsilon<h \\ \lambda_G(\varepsilon) & \text{for } h\le\varepsilon\le g \\ -\kappa_d(G) & \text{for } \varepsilon>g, \end{cases} \tag{3}$$

where λ_G is a piecewise continuously differentiable function on $[h,g]$ (g denotes the inradius throughout). Vol(G) is denoted again by $-\kappa_d(G)$.

For additional simplicity we also assume in our setting that there is a single generator G. (In the case of multiple generators we have to apply the formula to each generator separately and add them up.)

Remark 2.2. For any (not self-intersecting) polygon P in the plane, the volume of the inner ε-neighborhood of P is given by a quadratic polynomial in ε for sufficiently small ε. (A polygon has finitely many sides.) Let the polygon P have the vertices $A_1, A_2, \ldots, A_n$ (with $A_{n+1} = A_1$) in a certain successive ordering. Denote the length of the edge A_iA_{i+1} with a_i and the inner angle at A_i by α_i ($i = 1,2,\ldots,n$), see Fig.1. Then it can be shown that for small ε

$$V_P(\varepsilon) = \left(\sum_{i=1}^{n} a_i\right)\varepsilon - \left(\sum_{i=1}^{n}\delta_i\right)\varepsilon^2,$$

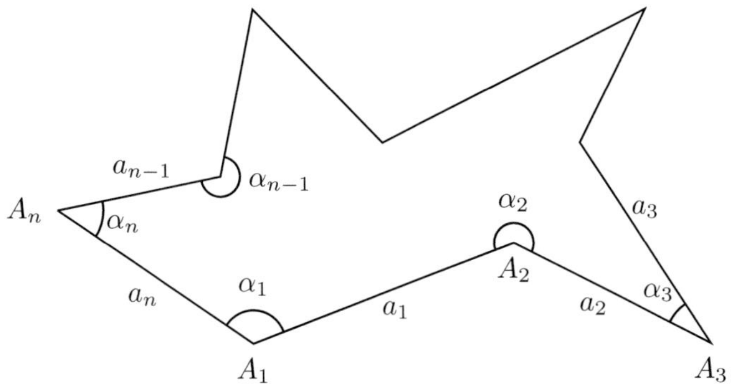

FIGURE 1. Polygon with n vertices.

where

$$\delta_i = \begin{cases} \cot \frac{\alpha_i}{2} & \text{for } 0 < \alpha_i < \pi \\ \frac{\pi}{2} - \frac{\alpha_i}{2} & \text{for } \pi \leq \alpha < 2\pi. \end{cases}$$

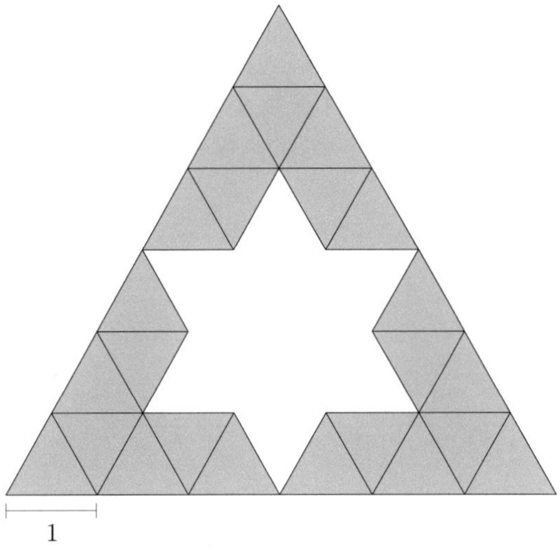

FIGURE 2. Maps of Example 2.3.

FIGURE 3. The attractor of Example 2.3.

Example 2.3. Consider the iterated function system $\Phi = \{\varphi_j\}_{j=1}^{24}$ on $\mathbb{R}^2$ with scaling ratios $\frac{1}{6}$ as indicated in Figure 2. The associated self-similar fractal is shown in Figure 3. This system satisfies the Pearse-Winter condition. The generator is a non-convex polygon G (with $a_i = 1$, $i = 1, 2, \ldots, 24$; $\alpha_i = \frac{\pi}{3}$ for odd i and $\alpha_i = \frac{4\pi}{3}$ for even i), see Figure 4. In Figure 5 the inner $\frac{1}{\sqrt{3}}$-neighborhood of G is shown.

By the formula above we have $V_G(\varepsilon) = 12\varepsilon - (6\sqrt{3} - \pi)\varepsilon^2$ for small ε. More precisely one can compute

$$V_G(\varepsilon) = \begin{cases} 12\varepsilon - (6\sqrt{3} - \pi)\varepsilon^2 & \text{for } 0 < \varepsilon < \frac{1}{\sqrt{3}} \\ \lambda_G(\varepsilon) & \text{for } \frac{1}{\sqrt{3}} \le \varepsilon \le 1 \\ 3\sqrt{3} & \text{for } \varepsilon > 1, \end{cases}$$

where $\lambda_G(\varepsilon) = \frac{3\sqrt{3}}{2} + \frac{3}{2}\sqrt{4\varepsilon^2 - 1} + (2\pi - 6\arccos\frac{1}{2\varepsilon})\varepsilon^2$. (Note that $\kappa_0(G) = \pi - 6\sqrt{3}$, $\kappa_1(G) = 12$ and $\kappa_2(G) = -3\sqrt{3}$.) It can be checked that $V_G(\varepsilon)$ is continuously differentiable on $[0, \infty)$ (see Figure 6).

We still use the notion of scaling ζ-function and the associated complex dimensions in the sense of L-P. Since the tubular ζ-function depends on the geometry of the generators, we want to define a new tubular ζ-function taking into account the type of generators satisfying (3).

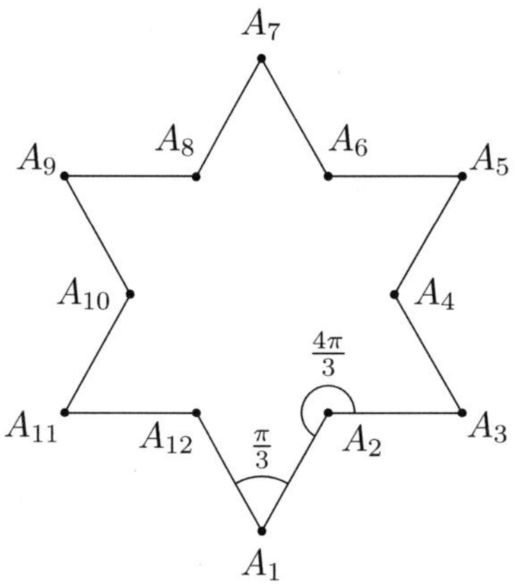

FIGURE 4. Generator G of Example 2.3.

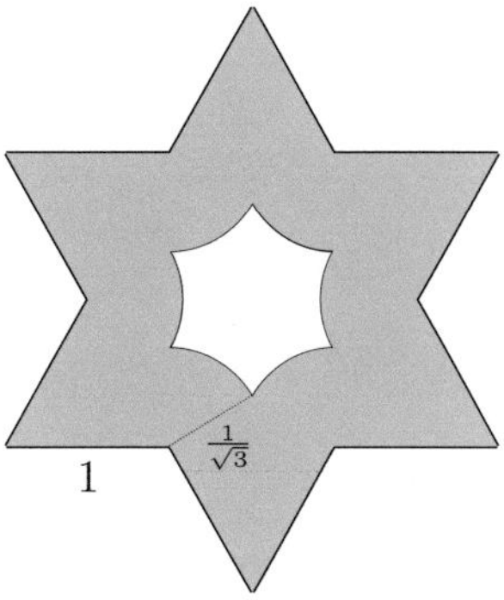

FIGURE 5. The inner $\varepsilon = \frac{1}{\sqrt{3}}$-neighborhood of G.

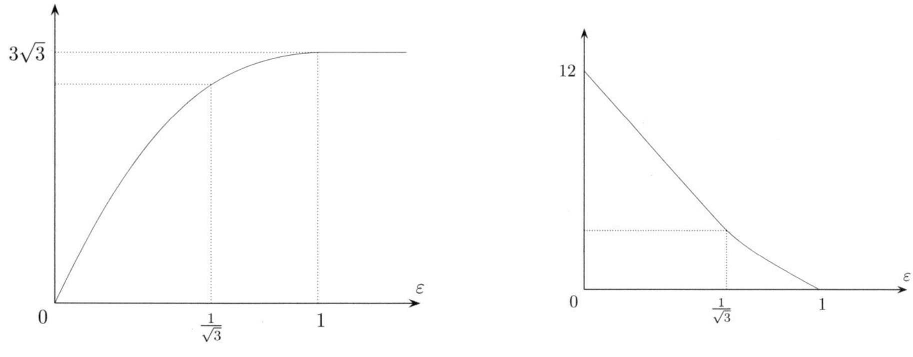

FIGURE 6. The graph of $V_G(\varepsilon)$ (left) and $V'_G(\varepsilon)$ (right).

Definition 2.4. The tubular ζ-function $\zeta_{\mathcal{T}}(s,\varepsilon)$ associated with the generator G satisfying (3) is defined by

$$\zeta_{\mathcal{T}}(s,\varepsilon) := \zeta(s)\varepsilon^{d-s}\left(\sum_{i=0}^{d-1}\frac{h^{s-i}}{s-i}\kappa_i(G) + \frac{g^{s-d}}{s-d}\kappa_d(G) + \Lambda(s)\right),$$

where $\Lambda(s)$ is an entire function given by

$$\Lambda(s) = \int_h^g u^{s-d-1}\lambda_G(u)du. \tag{4}$$

Now we can state our version of the L-P tube formula for generators satisfying condition (3):

Theorem 2.2 (Tube formula for tilings of self-similar fractals with non-Steiner-like generators).

$$V_{\mathcal{T}}(\varepsilon) = \sum_{\omega \in \mathfrak{D}_{\mathcal{T}}} \operatorname{res}(\zeta_{\mathcal{T}}(s,\varepsilon);\omega) \qquad \text{for } \varepsilon < h,$$

where $\mathfrak{D}_{\mathcal{T}} = \mathfrak{D} \cup \{0, 1, \ldots, d-1\}$.

Remark 2.4. We note that Theorem 2.2 (as well as Theorem 2.1) holds for the more general situation of [8] and even for more general sprays [4]. The specific Pearse-Winter condition (that the boundary of the convex hull of the fractal be a subset of the fractal) is not needed. But if this beautiful condition holds, then the tube formula yields a formula for the volume of the genuine ε-neighborhood of the fractal as explained in the introduction.

Remark 2.5. Note that d is not included in the set $\mathfrak{D}_{\mathcal{T}}$. The reason of this exclusion will be clear from the proof of the theorem.

Lemma 2.3. *Let the volume of the inner ε-neighborhood of a generator $G \subset \mathbb{R}^2$ be given as in (3):*

$$V_G(\varepsilon) = \begin{cases} \kappa_0\varepsilon^2 + \kappa_1\varepsilon & \text{for } 0 < \varepsilon < h \\ \lambda_G(\varepsilon) & \text{for } h \le \varepsilon \le g \\ -\kappa_2 & \text{for } \varepsilon > g. \end{cases}$$

We assume additionally $\lambda_G(\varepsilon) \in C^3[h,g]$. Then, using the notation

$$\begin{aligned} \Delta^i(x) &= \text{jump of the } i\text{-th derivative of } V_G \text{ at } x \\ &= V_G^i(x+) - V_G^i(x-), \end{aligned}$$

the tubular ζ-function

$$\zeta_{\mathcal{T}}(s,\varepsilon) = \zeta(s)\varepsilon^{2-s}\left(\frac{h^s}{s}\kappa_0 + \frac{h^{s-1}}{s-1}\kappa_1 + \frac{g^{s-2}}{s-2}\kappa_2 + \int_h^g u^{s-3}\lambda_G(u)du\right)$$

can be expressed as follows:

$$\begin{aligned} \zeta_{\mathcal{T}}(s,\varepsilon) &= \zeta(s)\varepsilon^{2-s}\left[\frac{1}{(s-1)(s-2)}\left(\Delta^1(h)h^{s-1} + \Delta^1(g)g^{s-1}\right)\right. \\ &\quad \left. - \frac{1}{s(s-1)(s-2)}\left(\Delta^2(h)h^s + \Delta^2(g)g^s + \int_h^g u^s\lambda_G'''(u)du\right)\right] \end{aligned}$$

Proof. Apply integration by parts. □

Remark 2.6. Similar formulas can be obtained if $\lambda_G(\varepsilon)$ is piecewise C^3 and also for higher dimensions.

Example 2.3 (continued). The scaling ζ-function of the system Φ of Example 2.3 is given by

$$\zeta(s) = \frac{1}{1 - \sum_{j=1}^{24} r_j^s} = \frac{1}{1 - 24(\frac{1}{6})^s}.$$

The set of complex dimensions is $\mathfrak{D} = \{D + \mathbb{i}np \mid n \in \mathbb{Z}\}$ where $D = \log_6 24$ and $p = \frac{2\pi}{\log 6}$.

The tubular ζ-function is given by

$$\zeta_{\mathcal{T}}(s,\varepsilon) = \zeta(s)\varepsilon^{2-s}\left(\frac{h^s}{s}(\pi - 6\sqrt{3}) + \frac{h^{s-1}}{s-1}12 + \frac{g^{s-2}}{s-2}(-3\sqrt{3}) + \Lambda(s)\right).$$

Applying the above lemma we obtain:

$$\zeta_{\mathcal{T}}(s,\varepsilon) = \zeta(s)\varepsilon^{2-s}\frac{-4\sqrt{3}}{s(s-1)(s-2)}\left(1 + \sqrt{3}\int_{\frac{1}{\sqrt{3}}}^{1}\frac{u^{s-1}}{(4u^2-1)^{3/2}}du\right).$$

Thus by Theorem 2.2 we get

$$\begin{aligned}
V_{\mathcal{T}}(\varepsilon) &= \sum_{\omega\in\mathfrak{D}\cup\{0,1\}} \operatorname{res}(\zeta_{\mathcal{T}}(s,\varepsilon);\omega) \\
&= \operatorname{res}(\zeta_{\mathcal{T}}(s,\varepsilon);0) + \operatorname{res}(\zeta_{\mathcal{T}}(s,\varepsilon);1) + \sum_{n\in\mathbb{Z}} \operatorname{res}(\zeta_{\mathcal{T}}(s,\varepsilon);D+\mathbb{i}np) \\
&= \frac{6\sqrt{3}-\pi}{23}\varepsilon^2 - 4\varepsilon \\
&\quad - \frac{4\sqrt{3}}{\log 6}\sum_{n\in\mathbb{Z}} \frac{\varepsilon^{2-D-\mathbb{i}np}}{(D+\mathbb{i}np)(D-1+\mathbb{i}np)(D-2+\mathbb{i}np)}\left(1+\sqrt{3}\int_{\frac{1}{\sqrt{3}}}^{1}\frac{u^{D-1+\mathbb{i}np}}{(4u^2-1)^{3/2}}du\right).
\end{aligned}$$

3. Proof of Theorem 2.2

Our goal is to find a closed expression for $V_{\mathcal{T}}(\varepsilon) = \sum_{n=1}^{\infty} V_{\mathcal{T}_n}(\varepsilon)$ as stated in Theorem 2.2. As we assumed a single generator for simplicity, the volume of the inner ε-tube of the tiling $\mathcal{T}$ is, by the tileset condition, the sum of the volumes of the inner ε-neighborhoods of all the scaled copies of G appearing in the tiling:

$$V_{\mathcal{T}}(\varepsilon) = \sum_{k=0}^{\infty}\sum_{w\in W_k} V_{r_w G}(\varepsilon),$$

where W_k and r_w are as in Definition 2.1 and $r_w G$ is a copy of G scaled by r_w. A simple calculation shows that, if $V_G(\varepsilon)$ is given as in (3), then

$$V_{xG}(\varepsilon) = \begin{cases} \sum_{j=0}^{d-1} \kappa_j(G) x^j \varepsilon^{d-j} & \text{for } \varepsilon < xh \\ x^d \lambda_G(\frac{\varepsilon}{x}) & \text{for } xh \leq \varepsilon \leq xg \\ -x^d \kappa_d(G) & \text{for } \varepsilon > xg. \end{cases}$$

It will be more convenient for us to regard $V_{xG}(\varepsilon)$ as a two-variable function of x and ε:

$$V_G(x,\varepsilon) := V_{xG}(\varepsilon) = \begin{cases} -x^d \kappa_d(G) & \text{for } 0 < x < \frac{\varepsilon}{g} \\ x^d \lambda_G(\frac{\varepsilon}{x}) & \text{for } \frac{\varepsilon}{g} \leq x \leq \frac{\varepsilon}{h} \\ \sum_{j=0}^{d-1} \kappa_j(G) x^j \varepsilon^{d-j} & \text{for } x > \frac{\varepsilon}{h}. \end{cases}$$

Recall that the Mellin transform $\mathcal{M}[f;s]$ of a function $f:(0,\infty) \to \mathbb{R}$ is given by

$$\mathcal{M}[f;s] = \widetilde{f}(s) = \int_0^\infty x^{s-1} f(x) dx.$$

For fixed ε, we take the Mellin transform of $V_G(x,\varepsilon)$ as a function of x:

$$\begin{aligned} \widetilde{V}_G(s,\varepsilon) &= \int_0^{\frac{\varepsilon}{g}} x^{s-1}(-x^d \kappa_d(G)) dx + \int_{\frac{\varepsilon}{g}}^{\frac{\varepsilon}{h}} x^{s-1} x^d \lambda_G(\frac{\varepsilon}{x}) dx \\ &\quad + \int_{\frac{\varepsilon}{h}}^{\infty} x^{s-1} \left(\sum_{j=0}^{d-1} \kappa_j(G) x^j \varepsilon^{d-j} \right) dx \\ &= -\varepsilon^{s+d} \left(\frac{g^{-s-d}}{s+d} \kappa_d(G) + \sum_{j=0}^{d-1} \frac{h^{-s-j}}{s+j} \kappa_j(G) \right) + \int_h^g \varepsilon^{s+d} u^{-s-d-1} \lambda_G(u) du, \end{aligned}$$

for $-d < \mathrm{Re}(s) < 1-d$. (In the second integral we made the change of variables $\frac{\varepsilon}{x} \to u$.)

$$\begin{aligned} \widetilde{V}_G(s,\varepsilon) &= -\varepsilon^{s+d} \left(\frac{g^{-s-d}}{s+d} \kappa_d(G) + \sum_{j=0}^{d-1} \frac{h^{-s-j}}{s+j} \kappa_j(G) - \int_h^g u^{-s-d-1} \lambda_G(u) du \right) \\ &= -\varepsilon^{s+d} \left(\frac{g^{-s-d}}{s+d} \kappa_d(G) + \sum_{j=0}^{d-1} \frac{h^{-s-j}}{s+j} \kappa_j(G) - \Lambda_G(-s) \right), \end{aligned} \tag{5}$$

where $\Lambda(s) = \int_h^g u^{s-d-1} \lambda_G(u) du$.

Since $V_G(x,\varepsilon)$ is piecewise continuously differentiable as a function of x and therefore of bounded variation, we can take the inverse Mellin transform of $\widetilde{V}_G(s,\varepsilon)$ to obtain ([9, Theorem 28, p.46])

$$V_G(x,\varepsilon) = \mathcal{M}^{-1}\left[\widetilde{V}_G(s,\varepsilon); x\right] = \frac{1}{2\pi \mathrm{i}}\int_{-c-\mathrm{i}\infty}^{-c+\mathrm{i}\infty} x^{-s}\widetilde{V}_G(s,\varepsilon)ds,$$

where $d-1<c<d$. For an additional purpose below, we choose c such that $D<c<d$. (Note that the non-triviality condition implies $D<d$, [8, Corollary 2.13].)

We shall now insert the above expression for $V_G(x,\varepsilon)$ into the sum

$$V_{\mathcal{T}}(\varepsilon) = \sum_{k=0}^{\infty}\sum_{w\in W_k} V_{r_w G}(\varepsilon),$$

but for the ease of the computation, we order the scaling coefficients r_w into a sequence $\{x_m\}_{m=1}^{\infty}$:

$$V_{\mathcal{T}}(\varepsilon) := \sum_{m=1}^{\infty} V_{x_m G}(\varepsilon),$$

so that we get

$$V_{\mathcal{T}}(\varepsilon) = \sum_{m=1}^{\infty}\frac{1}{2\pi \mathrm{i}}\int_{-c-\mathrm{i}\infty}^{-c+\mathrm{i}\infty} x_m^{-s}\widetilde{V}_G(s,\varepsilon)ds. \tag{6}$$

Changing the order of the sum and the integral (a detailed verification of this can be found in [3, p.12-15]), we obtain

$$\begin{aligned} V_{\mathcal{T}}(\varepsilon) &= \frac{1}{2\pi \mathrm{i}}\int_{-c-\mathrm{i}\infty}^{-c+\mathrm{i}\infty}\sum_{m=1}^{\infty} x_m^{-s}\widetilde{V}_G(s,\varepsilon)ds \\ &= \frac{1}{2\pi \mathrm{i}}\int_{-c-\mathrm{i}\infty}^{-c+\mathrm{i}\infty}\zeta(-s)\widetilde{V}_G(s,\varepsilon)ds \quad \text{by Def.2.1.} \end{aligned}$$

Changing the variable of the integral by $s' = -s$ we find

$$V_{\mathcal{T}}(\varepsilon) = \frac{1}{2\pi \mathrm{i}}\int_{c-\mathrm{i}\infty}^{c+\mathrm{i}\infty}\zeta(s)\widetilde{V}_G(-s,\varepsilon)ds.$$

By (5) we get

$$\begin{aligned} V_{\mathcal{T}}(\varepsilon) &= \frac{1}{2\pi \mathrm{i}}\int_{c-\mathrm{i}\infty}^{c+\mathrm{i}\infty}\zeta(s)\varepsilon^{d-s}\left(\frac{g^{s-d}}{s-d}\kappa_d(G) + \sum_{j=0}^{d-1}\frac{h^{s-j}}{s-j}\kappa_j(G) + \Lambda(s)\right)ds \\ &= \frac{1}{2\pi \mathrm{i}}\int_{c-\mathrm{i}\infty}^{c+\mathrm{i}\infty}\zeta_{\mathcal{T}}(s,\varepsilon)ds. \end{aligned}$$

The poles of $\zeta(s)$ are contained in a horizontally bounded strip $D_\ell \le \mathrm{Re}(s) \le D$, for some real number D_ℓ, which can be assumed to be negative also (see [4, Theorem. 2.17]).

Let c_ℓ be chosen such that

$$\text{i)}\quad c_\ell < D_\ell \tag{7}$$

$$\text{ii)}\quad \zeta(s) \text{ is bounded for } \mathrm{Re}(s) \leq c_\ell. \tag{8}$$

(The second property is possible because $|\zeta(s)| \to 0$ as $\mathrm{Re}(s) \to -\infty$.)

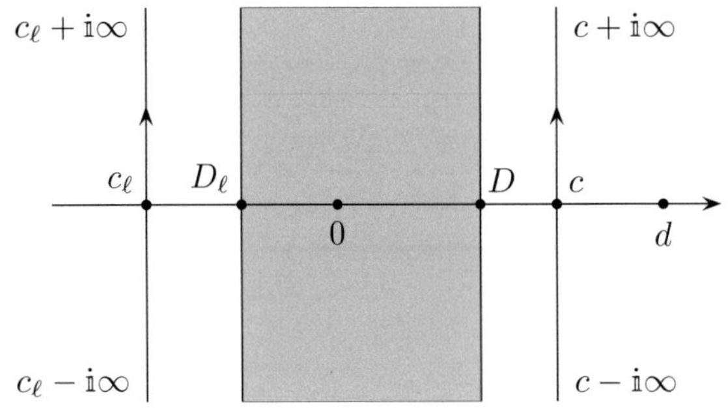

FIGURE 7. All the poles of $\zeta_\mathcal{T}$ are on the shaded strip.

We now proceed to show that (see Fig.7)

$$\frac{1}{2\pi\mathbb{i}}\int_{c-\mathbb{i}\infty}^{c+\mathbb{i}\infty} \zeta_\mathcal{T}(s,\varepsilon)ds = \frac{1}{2\pi\mathbb{i}}\int_{c_\ell-\mathbb{i}\infty}^{c_\ell+\mathbb{i}\infty} \zeta_\mathcal{T}(s,\varepsilon)ds + \sum_{\omega\in\mathfrak{D}\cup\{0,1,\ldots,d-1\}} \mathrm{res}(\zeta_\mathcal{T}(s,\varepsilon);\omega) \tag{9}$$

and

$$\frac{1}{2\pi\mathbb{i}}\int_{c_\ell-\mathbb{i}\infty}^{c_\ell+\mathbb{i}\infty} \zeta_\mathcal{T}(s,\varepsilon)ds = 0 \quad \text{for } \varepsilon < h. \tag{10}$$

Clearly, this will complete the proof of the theorem.

We begin with (9): First note that there exists a positive increasing sequence $\{\tau_n\}_{n=1}^\infty$ with $\{\tau_n\} \to \infty$ such that

$$|\zeta(\sigma \pm \mathbb{i}\tau_n)| \leq M \text{ for } c_\ell \leq \sigma \leq d \text{ and for all } n, \tag{11}$$

where M is some positive constant. (see [4, Theorem 3.26].)

Denote the rectangle with the corners $c - \mathbb{i}\tau_n, c + \mathbb{i}\tau_n, c_\ell + \mathbb{i}\tau_n, c_\ell - \mathbb{i}\tau_n$ (and with the edges $L_{1,n}, L_{2,n}, L_{3,n}, L_{4,n}$) by S_n (see Figure 8). By the residue theorem

$$\frac{1}{2\pi\mathbb{i}}\int_{\partial S_n} \zeta_\mathcal{T}(s,\varepsilon)ds = \sum_{\omega\in(\mathfrak{D}\cup\{0,1,\ldots,d-1\})\cap S_n} \mathrm{res}(\zeta_\mathcal{T}(s,\varepsilon);\omega).$$

(Note that d lies outside the rectangles S_n and does not contribute to the above sum.)

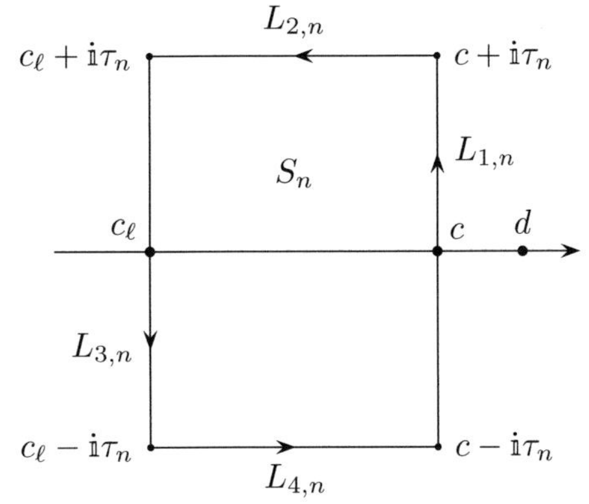

FIGURE 8. The rectangle S_n.

Therefore

$$\frac{1}{2\pi\mathbb{i}}\left(\int_{L_{1,n}}\zeta_{\mathcal{T}}(s,\varepsilon)ds+\int_{L_{2,n}}\zeta_{\mathcal{T}}(s,\varepsilon)ds+\int_{L_{3,n}}\zeta_{\mathcal{T}}(s,\varepsilon)ds+\int_{L_{4,n}}\zeta_{\mathcal{T}}(s,\varepsilon)ds\right) \tag{12}$$
$$=\sum_{\omega\in(\mathfrak{D}\cup\{0,1,\ldots,d-1\})\cap S_n}\operatorname{res}(\zeta_{\mathcal{T}}(s,\varepsilon);\omega).$$

Let us consider first the integral over $L_{2,n}$. For $s\in L_{2,n}$ we have $s=\sigma+\mathbb{i}\tau_n$, $c_\ell\le\sigma\le c$.

$$\begin{aligned}
|\zeta_{\mathcal{T}}(s,\varepsilon)| &= \left|\zeta(s)\varepsilon^{d-s}\left(\sum_{j=0}^{d-1}\frac{h^{s-j}}{s-j}\kappa_j(G)+\frac{g^{s-d}}{s-d}\kappa_d(G)+\Lambda(s)\right)\right|\\
&\le M\varepsilon^{d-\sigma}\left(\sum_{j=0}^{d-1}\frac{h^{\sigma-j}}{|\sigma+\mathbb{i}\tau_n-j|}|\kappa_j(G)|+\frac{g^{\sigma-d}}{|\sigma+\mathbb{i}\tau_n-d|}|\kappa_d(G)|+|\Lambda(s)|\right)\\
&\le M\varepsilon^{d-\sigma}\left(\sum_{j=0}^{d-1}\frac{h^{\sigma-j}|\kappa_j(G)|}{\tau_n}+\frac{g^{\sigma-d}|\kappa_d(G)|}{\tau_n}\right)+M\varepsilon^{d-\sigma}|\Lambda(s)|\\
&\le \frac{MM'}{\tau_n}+M\varepsilon^{d-\sigma}|\Lambda(s)|,
\end{aligned}$$

where $M'=\max_{c_\ell\le\sigma\le c}\left\{\varepsilon^{d-\sigma}\left(\sum_{j=0}^{d-1}h^{\sigma-j}|\kappa_j(G)|+g^{\sigma-d}|\kappa_d(G)|\right)\right\}$, and we used (11) to obtain the first inequality.

Now we consider the term $M\varepsilon^{d-\sigma}|\Lambda(s)|$: Recall that

$$\Lambda(s)=\int_h^g u^{s-d-1}\lambda_G(u)du.$$

Integrating by parts,

$$\Lambda(s) = \frac{u^{s-d}}{s-d}\lambda_G(u)\Big|_h^g - \int_h^g \frac{u^{s-d}}{s-d}\lambda'_G(u)du, \tag{13}$$

whence we can write $\varepsilon^{d-\sigma}|\Lambda(s)| \leq \frac{M''}{|s-d|} \leq \frac{M''}{\tau_n}$ by the assumption of continuity and piecewise continuous differentiability of $\lambda_G(u)$ and $c_\ell \leq \sigma \leq c$. Thus we obtain

$$|\zeta_\mathcal{T}(s,\varepsilon)| \leq \frac{MM'}{\tau_n} + \frac{MM''}{\tau_n}.$$

Therefore,

$$\left|\int_{L_{2,n}} \zeta_\mathcal{T}(s,\varepsilon)\, ds\right| \leq \frac{M(M'+M'')}{\tau_n}(c - c_\ell) \to 0 \text{ for } n \to \infty,$$

since $\tau_n \to \infty$ as $n \to \infty$.

Similarly the integral over $L_{4,n}$ tends to 0 as $n \to \infty$. Thus letting $n \to \infty$ in (12) gives (9).

Now, we will show (10): Let L_n be the line segment $L_n(t) = c_\ell + \mathbb{i}t$, $-n \leq t \leq n$; C_n be the semicircle $C_n(t) = c_\ell + ne^{\mathbb{i}t}$, $\frac{\pi}{2} \leq t \leq \frac{3\pi}{2}$ and $\Gamma_n = L_n + C_n$ (see Figure 9). By the choice of c_ℓ, $\zeta_\mathcal{T}(s,\varepsilon)$ is analytic on and inside Γ_n and therefore

$$\int_{L_n} \zeta_\mathcal{T}(s,\varepsilon)ds = -\int_{C_n} \zeta_\mathcal{T}(s,\varepsilon)ds.$$

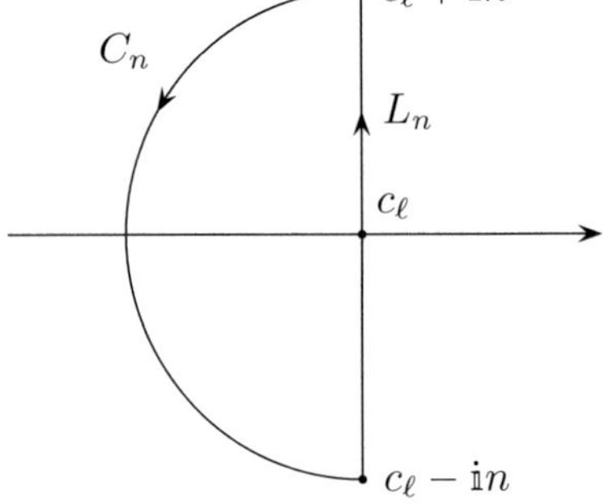

FIGURE 9. The contour $\Gamma_n = L_n + C_n$.

We will be done when we show that the right-hand side tends to 0 as $n \to \infty$. Using the parametrization of C_n, we obtain

$$\left|\int_{C_n} \zeta_{\mathcal{T}}(s,\varepsilon)ds\right| = \left|\int_{\frac{\pi}{2}}^{\frac{3\pi}{2}} \zeta(c_\ell + ne^{\mathrm{i}t})\varepsilon^{d-c_\ell-ne^{\mathrm{i}t}} \left(\sum_{j=0}^{d-1} \frac{h^{c_\ell+ne^{\mathrm{i}t}-j}}{c_\ell+ne^{\mathrm{i}t}-j}\kappa_j(G) + \frac{g^{c_\ell+ne^{\mathrm{i}t}-d}}{c_\ell+ne^{\mathrm{i}t}-d}\kappa_d(G) + \Lambda(c_\ell+ne^{\mathrm{i}t})\right) \mathrm{i}ne^{\mathrm{i}t}dt\right|.$$

By condition (8), $\zeta(s)$ is bounded for $\mathrm{Re}(s) \leq c_\ell$, say $|\zeta(s)| \leq K$:

$$\begin{aligned}\left|\int_{C_n} \zeta_{\mathcal{T}}(s,\varepsilon)ds\right| \leq\ & K\varepsilon^{d-c_\ell}\left(\sum_{j=0}^{d-1} \frac{n\,h^{c_\ell-j}}{n-|c_\ell-j|}|\kappa_j(G)| \int_{\frac{\pi}{2}}^{\frac{3\pi}{2}} \left(\frac{h}{\varepsilon}\right)^{n\cos t} dt\right. \\ & + \frac{n\,g^{c_\ell-d}}{n-|c_\ell-d|}|\kappa_d(G)| \int_{\frac{\pi}{2}}^{\frac{3\pi}{2}} \left(\frac{g}{\varepsilon}\right)^{n\cos t} dt \\ & \left. + n\int_{\frac{\pi}{2}}^{\frac{3\pi}{2}} \left|\Lambda(c_\ell+ne^{\mathrm{i}t})\right| \varepsilon^{-n\cos t}dt\right).\end{aligned}$$

Let us denote the right-hand side by $I_1+I_2+I_3$. We have

$$I_1 = K\varepsilon^{d-c_\ell}\sum_{j=0}^{d-1} \frac{n\,h^{c_\ell-j}}{n-|c_\ell-j|}|\kappa_j(G)| \int_{\frac{\pi}{2}}^{\frac{3\pi}{2}} \left(\frac{h}{\varepsilon}\right)^{n\cos t} dt \leq K_1 \int_{\frac{\pi}{2}}^{\frac{3\pi}{2}} \left(\frac{h}{\varepsilon}\right)^{n\cos t} dt,$$

where K_1 is some constant.

The well-known Jordan Lemma [1] states that

$$\lim_{n\to\infty} \int_{\frac{\pi}{2}}^{\frac{3\pi}{2}} a^{n\cos t}dt = 0,$$

for any fixed $a > 1$. Since we assumed $\varepsilon < h$ for Theorem 2.2, we thus get $I_1 \to 0$ as $n \to \infty$.

Similarly $I_2 \to 0$ as $n \to \infty$, since $\varepsilon < g$ as $h < g$.

To deal with the term I_3, recall that by (13)

$$\Lambda(s) = \frac{u^{s-d}}{s-d}\lambda_G(u)\Big|_h^g - \int_h^g \frac{u^{s-d}}{s-d}\lambda_G'(u)du.$$

By continuity and piecewise continuous differentiability of λ_G, we can write

$$|\Lambda(s)| \leq \frac{1}{|s-d|}\left(K_1\,g^{\mathrm{Re}(s)-d} + K_2\,h^{\mathrm{Re}(s)-d}\right) + \frac{1}{|s-d|}\int_h^g u^{\mathrm{Re}(s)-d} K_3\,du$$

with some constants K_1, K_2, K_3.

For $s = c_\ell + ne^{\mathrm{i}t}, (\frac{\pi}{2} \le t \le \frac{3\pi}{2})$ we have $\mathrm{Re}(s) = c_\ell + n\cos t < 0$ and $u^{\mathrm{Re}(s)-d}$ is a decreasing function on $[h, g]$. Thus,

$$\begin{aligned} |\Lambda(s)| &\le K_4 \frac{1}{|s-d|} h^{\mathrm{Re}(s)-d} + K_3 \frac{1}{|s-d|}(g-h)h^{\mathrm{Re}(s)-d} \\ &\le K_5 \frac{1}{|c_\ell + ne^{\mathrm{i}t} - d|} h^{c_\ell + n\cos t - d} \\ &\le K_6 \frac{h^{n\cos t}}{n - |c_\ell - d|}. \end{aligned} \tag{14}$$

Now we show that $I_3 \to 0$ as $n \to \infty$. Since

$$I_3 = n \int_{\frac{\pi}{2}}^{\frac{3\pi}{2}} \left|\Lambda\left(c_\ell + ne^{\mathrm{i}t}\right)\right| \varepsilon^{-n\cos t} dt,$$

using (14)

$$I_3 \le \frac{K_6\, n}{n - |c_\ell - d|} \int_{\frac{\pi}{2}}^{\frac{3\pi}{2}} \left(\frac{h}{\varepsilon}\right)^{n\cos t} dt \to 0 \quad \text{as } n \to \infty,$$

by Jordan Lemma. Hence the claim (10) is verified and thus the proof of Theorem 2.2 is completed.

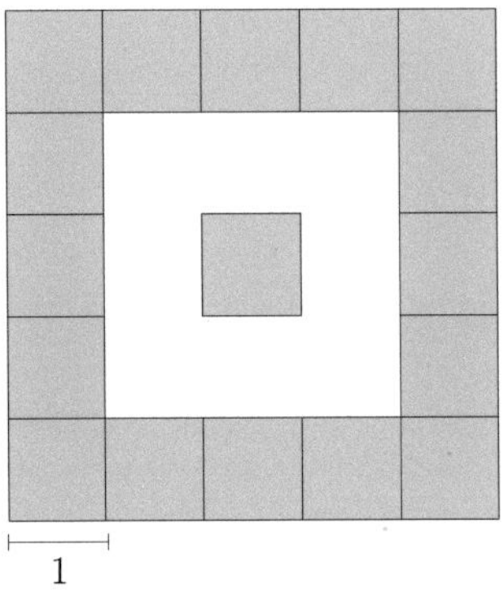

FIGURE 10. Maps of Example 4.1.

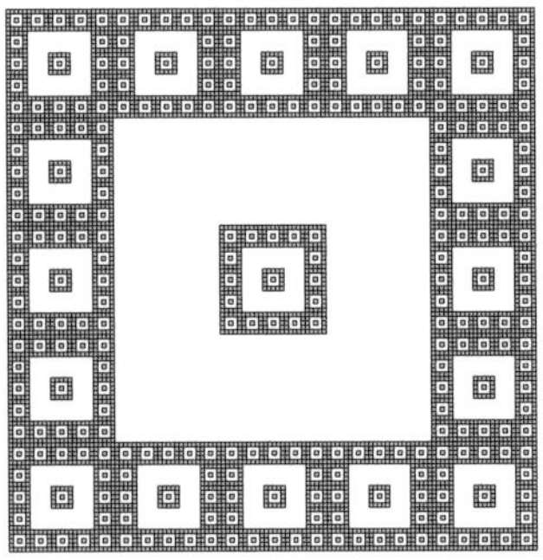

FIGURE 11. The attractor of Example 4.1.

4. Further Examples

4.1. An example with non-simply connected generator

Consider the iterated function system $\Phi = \{\varphi_j\}_{j=1}^{17}$ on $\mathbb{R}^2$ with scaling ratios $\frac{1}{5}$ as indicated in Figure 10. The associated self-similar fractal is shown in Figure 11. This

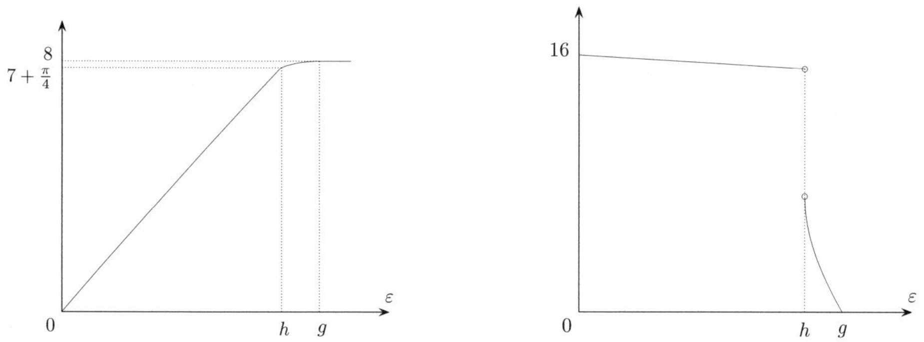

FIGURE 12. The graph of $V_G(\varepsilon)$ (left) and $V_G'(\varepsilon)$ (right) of the generator G of Example 4.1.

system satisfies the Pearse-Winter condition. The generator G is a non-simply connected set, where the volume of the inner ε-tube is given by the following function:

$$V_G(\varepsilon) = \begin{cases} 16\varepsilon + (\pi-4)\varepsilon^2 & \text{for} \quad 0 < \varepsilon < h \\ \lambda_G(\varepsilon) & \text{for} \quad h \le \varepsilon \le g \\ 8 & \text{for} \quad \varepsilon > g, \end{cases}$$

where

$$\lambda_G(\varepsilon) = 4 + 8\varepsilon + (\pi-4)\varepsilon^2 + 4(1-\varepsilon)\sqrt{2\varepsilon-1} - 4\varepsilon^2 \arctan\left(\frac{\sqrt{2\varepsilon-1}}{1-\varepsilon}\right),$$

and $h = \frac{1}{2}$, $g = 2-\sqrt{2}$. Note that $\kappa_0(G) = \pi - 4$, $\kappa_1(G) = 16$, $\kappa_2(G) = -8$. We remark that the function $V_G(\varepsilon)$ is not differentiable (at $\varepsilon = \frac{1}{2}$) (see Figure 12). The scaling ζ-function of the system Φ is given by

$$\zeta(s) = \frac{1}{1 - \sum_{j=1}^{17} r_j^s} = \frac{1}{1 - 17\left(\frac{1}{5}\right)^s}.$$

The set of complex dimensions is $\mathfrak{D} = \{D + \mathbb{i}np \mid n \in \mathbb{Z}\}$, where $D = \log_5 17$ and $p = \frac{2\pi}{\log 5}$. The tubular ζ-function is given by

$$\zeta_{\mathcal{T}}(s,\varepsilon) = \zeta(s)\varepsilon^{2-s}\left(\frac{h^s}{s}(\pi-4) + \frac{h^{s-1}}{s-1}16 + \frac{g^{s-2}}{s-2}(-8) + \Lambda(s)\right),$$

where $\Lambda(s)$ is as in (4).

Applying Theorem 2.2 we get

$$\begin{aligned} V_{\mathcal{T}}(\varepsilon) &= \sum_{\omega \in \mathfrak{D} \cup \{0,1\}} \operatorname{res}(\zeta_{\mathcal{T}}(s,\varepsilon);\omega) \\ &= \operatorname{res}(\zeta_{\mathcal{T}}(s,\varepsilon);0) + \operatorname{res}(\zeta_{\mathcal{T}}(s,\varepsilon);1) + \sum_{n\in\mathbb{Z}} \operatorname{res}(\zeta_{\mathcal{T}}(s,\varepsilon); D + \mathbb{i}np) \\ &= \frac{4-\pi}{16}\varepsilon^2 - \frac{20}{3}\varepsilon + \frac{1}{\log 5}\sum_{n\in\mathbb{Z}} \varepsilon^{2-D-\mathbb{i}np}\left(\frac{h^{D+\mathbb{i}np}}{D+\mathbb{i}np}(\pi-4) + \frac{h^{D-1-\mathbb{i}np}}{D-1-\mathbb{i}np}16 + \right. \\ &\qquad \left. + \frac{g^{D-2-\mathbb{i}np}}{D-2-\mathbb{i}np}(-8) + \Lambda(D+\mathbb{i}np)\right). \end{aligned}$$

4.2. A non-lattice example

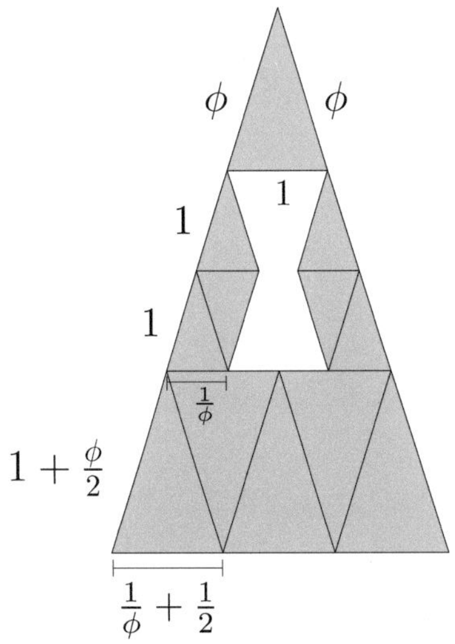

FIGURE 13. Maps of Example 4.2.

FIGURE 14. The attractor of Example 4.2.

A self-similar fractal F with scaling ratios r_j, $(j = 1,\dots,J)$ is called of lattice type if the additive subgroup $\sum_{j=1}^{J}(\log r_j)\mathbb{Z} \subset \mathbb{R}$ is discrete and otherwise of non-lattice type.

Consider the iterated function system $\Psi = \{\psi\}_{j=1}^{12}$ on $\mathbb{R}^2$ with scaling ratios

$$r_1 = \frac{1}{3\left(\frac{1}{\phi}+\frac{1}{2}\right)} = \frac{2\phi}{3\phi+6},$$

$$r_2 = \frac{\frac{1}{\phi}}{3\left(\frac{1}{\phi}+\frac{1}{2}\right)} = \frac{2}{3\phi+6}, \qquad (\phi \text{ being the golden ratio})$$

and $r_3 = \frac{1}{3}$ as indicated in Figure 13. The associated self-similar fractal is shown in Figure 14. This system satisfies the Pearse-Winter condition. The generator is a non-convex set, with the following volume function for the inner ε-tube:

$$V_G(\varepsilon) = \begin{cases} 6\varepsilon - \left(\frac{4\sqrt{4\phi+3}}{2\phi-1} - \frac{\pi}{5}\right)\varepsilon^2 & \text{for} \quad 0 < \varepsilon < h \\ \lambda_G(\varepsilon) & \text{for} \quad h \le \varepsilon \le g \\ \frac{1}{2}\frac{(2\phi-1)\sqrt{4\phi+3}}{\phi+1} & \text{for} \quad \varepsilon > g, \end{cases}$$

where

$$\lambda_G(\varepsilon) = \begin{cases} 6\varepsilon - \left(\frac{4\sqrt{4\phi+3}}{2\phi-1} - \frac{\pi}{5}\right)\varepsilon^2 - 2\varepsilon^2 \arccos\left(\frac{\phi-1}{2\phi\varepsilon}\right) + \frac{\phi-1}{\phi}\sqrt{\varepsilon^2 - \frac{2-\phi}{4\phi+4}}, & \text{if } h \le \varepsilon < h' \\ \frac{1}{2}\frac{(\phi-2)\sqrt{4\phi+3}}{\phi+1} + 2\varepsilon\sqrt{2\phi+1} - 2\varepsilon^2\frac{8\phi+5}{\sqrt{4\phi+3}}, & \text{if } h' \le \varepsilon \le g, \end{cases}$$

and $h = \frac{\phi-1}{2\phi}$, $h' = \frac{\phi-1}{\sqrt{4\phi+3}}$ and $g = \frac{\sqrt{4\phi+3}}{4\phi+2}$.

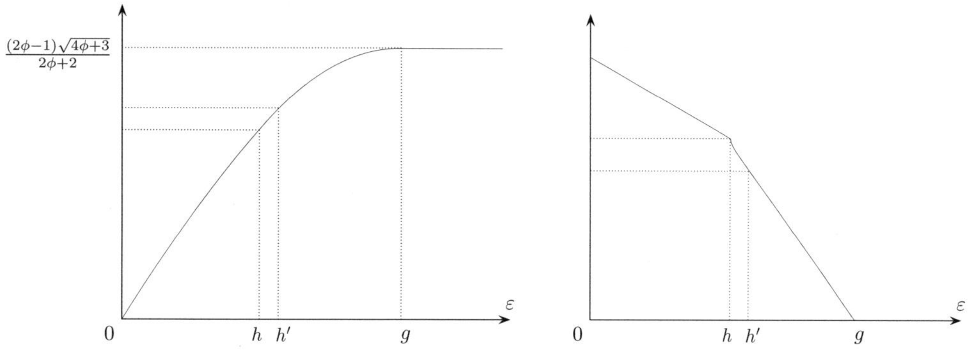

FIGURE 15. The graph of $V_G(\varepsilon)$ (left) and $V_G'(\varepsilon)$ (right) of the generator G of Example 4.2.

We remark that the function $V_G(\varepsilon)$ is differentiable (see Figure 15). The scaling ζ-function of the system is given by

$$\zeta(s) = \frac{1}{1 - \left(\frac{2\phi}{3\phi+6}\right)^s - 6\left(\frac{2}{3\phi+6}\right)^s - 5\left(\frac{1}{3}\right)^s}.$$

The dimension D of this example is approximately 1.8689. As this example is of non-lattice type, we can not exactly compute the complex dimensions [4, p.66], but we give several approximations of the set of the complex dimensions in Figure 17 along the lines of computation of [4].

The tubular ζ-function is given by

$$\zeta_{\mathcal{T}}(s,\varepsilon) = \zeta(s)\varepsilon^{2-s}\left(\frac{h^s}{s}\left(\frac{4\sqrt{4\phi+3}}{2\phi-1} - \frac{\pi}{5}\right) + \frac{h^{s-1}}{s-1}6 + \frac{g^{s-2}}{s-2}k + \Lambda(s)\right),$$

where $k = \frac{1}{2}\frac{(2\phi-1)\sqrt{4\phi+3}}{\phi+1}$ and $\Lambda(s)$ is as in (4).

Appendix

To determine the complex dimensions of a self-similar fractal, we have to solve the Moran equation in the complex domain. As it is not possible to solve this equation in the non-lattice case exactly, we can only approximate solutions. We want to explain briefly the approach we used to plot the approximate complex dimensions for the last example. The Moran equation in this case is

$$r_1^s + 6r_2^s + 5r_3^s = 1,$$

where $r_1 = \frac{2\phi}{3\phi+6}$, $r_2 = \frac{2}{3\phi+6}$, $r_3 = \frac{1}{3}$.

We can rewrite this equation as follows:

$$\begin{aligned} e^{s\log r_1} + 6e^{s\log r_2} + 5e^{s\log r_3} &= 1 \\ e^{-s\log 3(-\log_3 r_1)} + 6e^{-s\log 3(-\log_3 r_2)} + 5e^{-s\log 3} &= 1. \end{aligned}$$

Now, we choose some appropriate rational approximations to $\log_3 r_1$ and $\log_3 r_2$, then solve the resulting polynomial equation and find approximate complex dimensions. We used for the pair $(-\log_3 r_1, -\log_3 r_2)$ the following pairs of approximations:

$$\left(\frac{10}{9},\frac{3}{2}\right), \left(\frac{11}{10},\frac{17}{11}\right), \left(\frac{65}{59},\frac{20}{13}\right), \left(\frac{76}{69},\frac{97}{63}\right).$$

To detail the first case, the pair $\left(\frac{10}{9},\frac{3}{2}\right)$ leads to the equation

$$e^{-(s\log 3)\left(\frac{10}{9}\right)} + 6e^{-s\log 3\left(\frac{3}{2}\right)} + 5e^{-s\log 3} = 1.$$

Setting $z = e^{-\frac{s}{18}\log 3}$, we get the polynomial equation

$$z^{20} + 6z^{27} + 5z^{18} = 1.$$

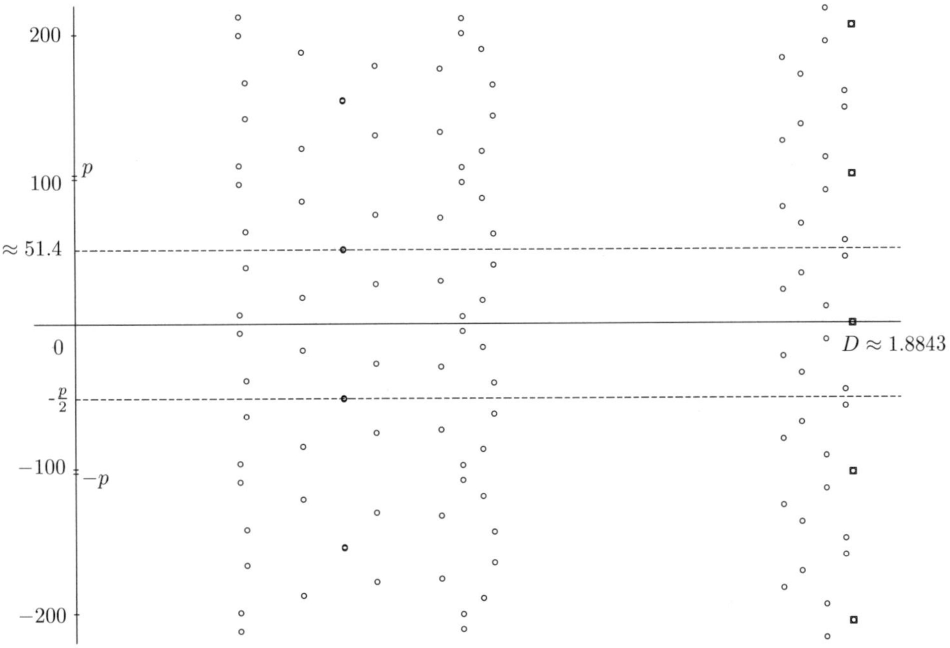

FIGURE 16. Approximate complex dimensions of Example 4.2 for $(-\log_3 r_1, -\log_3 r_2) \approx \left(\frac{10}{9}, \frac{3}{2}\right)$.

We solved this equation using Maple. There are 27 distinct roots, one of which is real. In terms of these roots, we can express the approximate complex dimensions as

$$s = -\frac{18 \log z}{\log 3} + \frac{18}{\log 3} 2\pi \mathrm{i} n, \qquad n \in \mathbb{Z}.$$

The real root $z \approx 0.8913$ gives for $n = 0$ the approximate similarity dimension $D \approx 1.8843$.

The pairwise-conjugate 13 root pairs give rise to 13 vertical lines, so that all approximate complex dimensions are located on 14 vertical lines with a period $\frac{18}{\log 3} 2\pi \mathrm{i}$ (see Figure 16).

The other approximations lead to the polynomials

$$z^{121} + 6z^{170} + 5z^{110} = 1$$
$$z^{845} + 6z^{1180} + 5z^{767} = 1$$
$$z^{4788} + 6z^{6693} + 5z^{4347} = 1$$

and the corresponding approximate complex dimensions are plotted in Figure 17.

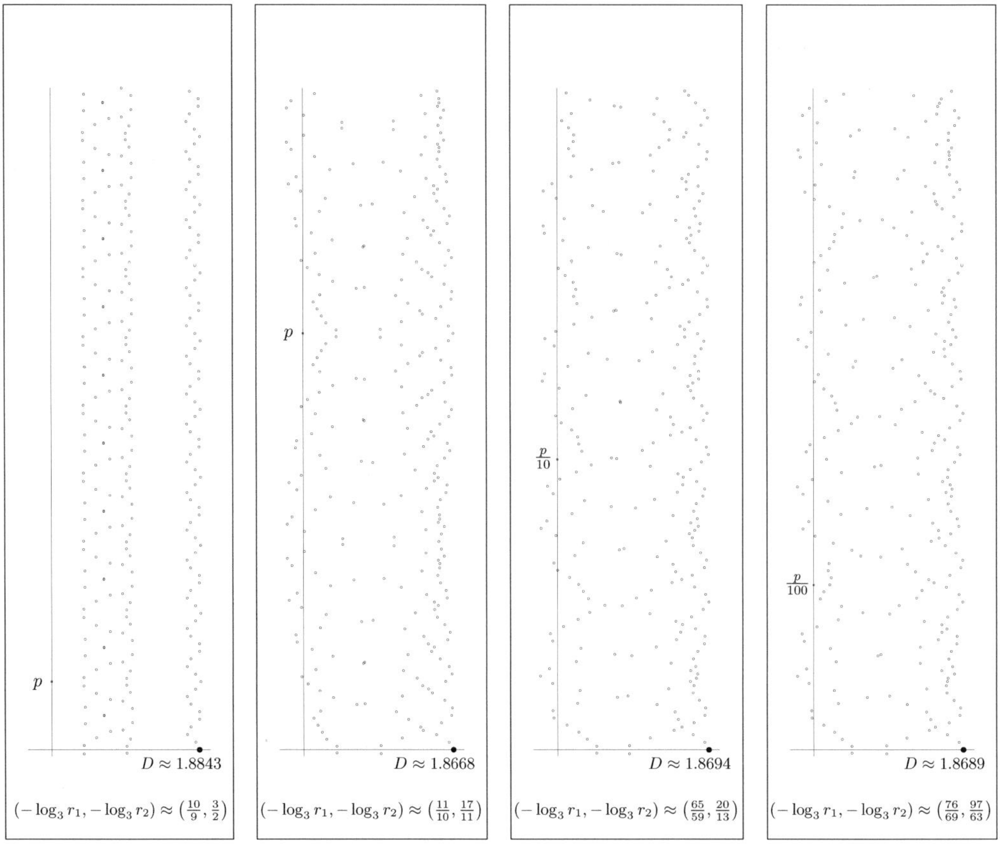

FIGURE 17. Consecutive approximations of complex dimensions of Example 4.2.

References

[1] R. V. Churchill, J. W. Brown, R. F. Verhey, *Complex variables and applications*, 3rd edition, McGraw-Hill, 1976.

[2] B. Demir, A. Deniz, Ş. Koçak, A. E. Üreyen, *Tube formulas for graph-directed fractals*, Fractals **18** (3), (2010), 349–361.

[3] A.Deniz, Ş. Koçak, Y. Özdemir and A. E. Üreyen, *Tube formula for self-similar fractals with non-Steiner-like generators*, preprint (2009), available at arXiv:math.MG/0911.4966.

[4] M. L. Lapidus and M. van Frankenhuijsen, *Fractal geometry, complex dimensions and zeta functions: geometry and spectra of fractal strings*, Springer-Verlag, New York, 2006.

[5] M. L. Lapidus and E. P. J. Pearse, *Tube formulas and complex dimensions of self-similar tilings*, Acta Appl. Math. **112**, (2010), 91–136.

[6] M. L. Lapidus, E. P. J. Pearse and S. Winter, *Pointwise tube formulas for fractal sprays and self-similar tilings with arbitrary generators*, Adv. Math. **227**, (2011), 1349–1398.

[7] E. P. J. Pearse, *Canonical self-affine tilings by iterated function systems*, Indiana Univ. Math J. **56** (6), (2007), 3151–3170.

[8] E.P.J. Pearse and S. Winter, *Geometry of canonical self-similar tilings*, Rocky Mountain J. Math. **42**(4) (2012), 1327-1357.

[9] E.C. Titchmarsh, *Introduction to the theory of Fourier integrals*, Oxford at the Clerandon Press, 1948.

Anadolu University, Department of Mathematics, 26470, Eskişehir, Turkey
E-mail address: adeniz@anadolu.edu.tr

Anadolu University, Department of Mathematics, 26470, Eskişehir, Turkey
E-mail address: skocak@anadolu.edu.tr

Anadolu University, Department of Mathematics, 26470, Eskişehir, Turkey
E-mail address: yunuso@anadolu.edu.tr

Anadolu University, Department of Mathematics, 26470, Eskişehir, Turkey
E-mail address: aeureyen@anadolu.edu.tr

Proceedings of 19th Gökova
Geometry-Topology Conference
pp. 146 – 153

Crowell's state space is connected

Daniel Selahi Durusoy

ABSTRACT. We study the set of Crowell states for alternating knot projections and show that for prime alternating knots the space of states for a reduced projection is connected, a result similar to that for Kauffman states. As an application we give a new proof of a result of Ozsváth and Szabó characterizing $(2, 2n + 1)$ torus knots among alternating knots.

1. Introduction

Some of the many definitions of the Alexander polynomial of a knot is stated through state sums. Kauffman has described and studied a state sum model for the Alexander polynomial in great detail [Kau83]. In an earlier paper Crowell has described another state sum model for the Alexander polynomial for the subclass of alternating knots ([Cro59], Theorem 2.12).

In the next section we will recall the definition of Crowell states and examine some of their properties. In Section 3 we will prove that

Theorem 1.1. *If K is an alternating prime knot and D is a reduced knot diagram for K, then any two states differ by a finite sequence of terminal edge exchanges.*

This theorem is similar in nature to the Clock Theorem of Kauffman [Kau83] which states that any two Kauffman states differ by a finite sequence of clockwise and counter-clockwise moves, which was also proven in the language of graphs in section 4 of [GL86]. This work is independent of those mentioned because of the simple reason that Kauffman states and Crowell states do not correspond to each other in any natural way as observed from the fact that the space of Crowell states do not form a lattice in general (see Proposition 2.6).

In section 4, as an application we will give an alternative proof that $(2, 2n + 1)$ torus knots are characterized by their Alexander polynomials among alternating knots, which was originally proven by Ozsváth and Szabó (Proposition 4.1 in [OS05]).

2. A state model

In this section we will review the definition of the state sum model for alternating links given by Crowell [Cro59] and investigate some properties of the states.

Key words and phrases. State sum, Alexander polynomial, spanning trees.

Given a knot K and an oriented alternating diagram D of K with n crossings we obtain a weighted labeled directed planar graph $G(D)$ as follows: replace a small neighborhood of each crossing by a degree 4 vertex according to the following figure (k is the vertex label):

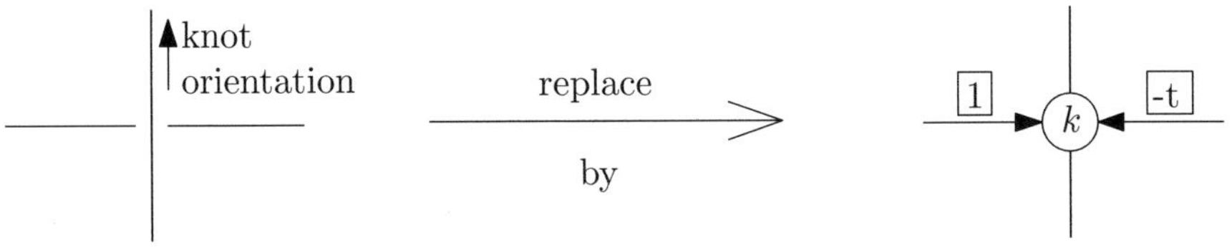

FIGURE 1. From a knot diagram to a directed graph.

Proposition 2.1. *This definition orients all edges and these orientations are compatible with orientations coming from a checkerboard coloring of the regions in the complement of D.*

Proof. Each edge gets an orientation since while traveling around a region the strands traveled alternate between under strand and over strand. Hence at each crossing if an edge is coming in, the boundary of the region will continue along the over strand, which becomes an under strand at its other end, hence we get an orientation on that edge as well, consistent with the previous edge. These orientations are compatible with a checkerboard coloring since crossing a region to another across an edge we get opposite orientations in the plane. □

Choose a vertex v_0 of $G(D)$, called the root. Directed paths starting at v_0 that don't contain cycles will be called rooted simple paths. Trees with edges oriented away from v_0 will be called rooted trees. Spanning rooted trees will be called states. Let $Tr(v_0)$ be the space of states and $w(T)$ be the product of weights of all edges in a state T. According to [Cro59, Theorem 2.12] we get the renormalized Alexander polynomial as a sum of monomials corresponding to each state by

$$\Delta_K(t) = (-t)^m \cdot \sum_{T \in Tr(v_0)} w(T) \qquad (1)$$

where m is chosen so that the term with the least power of t is a positive constant.

Proposition 2.2. *For any vertex v, there is a rooted simple path from v_0 to v in $G(D)$.*

Proof. Since K is a knot, $G(D)$ as an unoriented graph is connected. Pick an unoriented path $e_1, e_2, \ldots, e_m$ starting at v_0, ending at v. If e_i is not oriented away from the root, pick edges $e_{i,1}, e_{i,2}, \ldots e_{i,m_i}$ that go around one of the two regions adjacent to e_i. Due to Proposition 2.1, we get compatible orientations on this sequence of edges. At the end we get a rooted path which might visit some vertices more than once. For each vertex that is visited more than once, remove all edges between the first and last visits. □

Corollary 2.3. *Any rooted tree T can be extended to a rooted spanning tree $\tilde{T}$.*

Proof. For any vertex v not in T, find a rooted simple path α from v_0 to v using Proposition 2.2. Add enough of the final segment of α to the current tree so that the union will be connected without creating cycles. □

Proposition 2.4. *Given a reduced alternating diagram for a prime knot, any edge in $G(D)$ except those ending at v_0 appear as a teminal edge in at least one state.*

Proof. Call the given edge e_0 starting at vertex w_0, ending at vertex w_1. We will find a rooted simple path from v_0 to w_0, add the edge e_0, and extend this path into a rooted spanning tree.

Use Proposition 2.2 to construct a rooted simple path Γ from v_0 to w_0. There are two possible cases:
(a) Γ doesn't go through w_1: Do nothing extra.
(b) Γ goes through w_1 before reaching w_0: Adding e_0 to Γ produces a cycle. To avoid this problem we will go around as follows. Assume Γ contains an edge e' going from v' to w_1 and an edge e'' going from w_1 to v''. We need to connect v' to v'' by a directed simple path avoiding e' and e''. To achieve this, let R be the region bounded on two sides by e' and e'' and $\hat{R}$ be the union of regions adjacent to R along edges other than e' and e''. Then starting at v', following edges on the boundary of $\hat{R}$ that are not on the boundary of R, we get a directed path α ending at v''. Since K is prime and D is reduced, this new path α does not include e' and e'' since otherwise we could draw a separating circle passing through the e' or e'' and another common edge of R and $\hat{R}$. Replacing e' and e'' by α in Γ, we get a rooted path from v_0 to w_0 avoiding w_1. It could include cycles, which can be eliminated as in the proof of Proposition 2.2.

Next we need to extend the rooted simple path $\Gamma \cup e_0$ to a rooted spanning tree, keeping e_0 a terminal edge. Consider the two edges coming out from w_1, call them e_1' and e_1'', with terminal vertices w_1' and w_1''. Using a similar argument as in the proof of case (b) above, consider R being the region bounded by e_0 and e_1', and $\hat{R}$ the union of regions adjacent to R except along e_0 and e_1', and removing cycles we get a directed path β from w_0 to w_1' avoiding w_1. Starting at w_1', add enough edges from the final segment of β so that w_1' is connected to a vertex in $\Gamma \cup e_0$ without creating a cycle. Do similarly for w_1''.

Now, use Corollary 2.3 to extend this rooted tree to a rooted spanning tree. During this process e_0 stays a terminal edge since adding e_1' or e_1'' would create a cycle. □

The state sum in Equation 1 resembles the state sum defined by Kauffman [Kau83]. Kauffman has studied an operation called a *clock move* (transposition) that transforms a state to another that differs only at two crossings and showed that all states differ from one another by a sequence of clock moves. With that in mind we define the following operation for reduced alternating diagrams:

Definition 2.1. A state T_2 is obtained from a state T_1 by a *terminal edge exchange* move (edge exchange for short) if replacing a terminal edge in T_1 by the other incoming edge at the terminal vertex gives T_2.

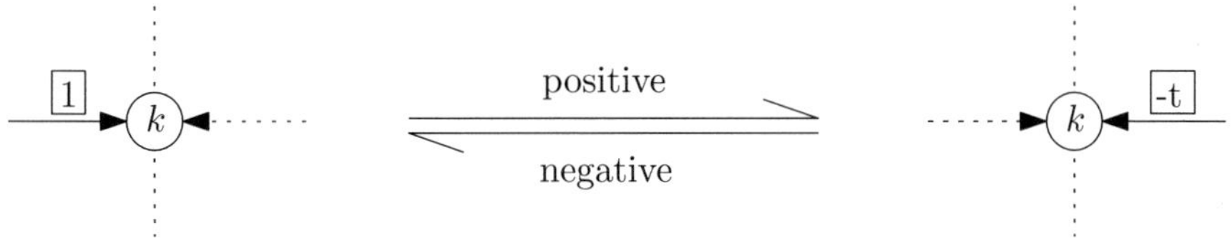

FIGURE 2. Terminal edge exchange.

Proposition 2.5. *At any terminal edge, an edge exchange gives a new state, except at edges ending at a kink.*

Proof. If there is a kink at v, there is a unique edge that connects to v in any spanning tree, since the opposite edge is a loop. At any other vertex, it is easy to check that one still gets a rooted spanning tree. □

Edge exchange gives a partial order on the set of states by defining the covering relation of the partial order as T_1 comes immediately before T_2 if T_2 is obtained from T_1 by one positive edge exchange.

Comparing these states with the black trees for Kauffman states, even though states are rooted spanning trees in both models, in Kauffman states the orientations on the edges are chosen after a spanning tree of the black graph is chosen, so the same edge can inherit different orientations in different states. Furthermore, consider the graph whose vertices are Crowell states and any two vertices are connected by an edge if there is a terminal edge exchange that takes one state to the other. The following proposition shows that edge exchanges do not correspond to clock moves under any bijection between the Crowell and Kauffman states since Kauffman states form a distributive lattice [Kau83].

Proposition 2.6. *The space of Crowell states is not a lattice in general with any choice of a partial order compatible with terminal edge exchanges even for reduced alternating diagrams for prime alternating knots.*

Proof. Figure 3 illustrates the graph of Crowell states for the knot 7_6 in Rolfsen's table. Let us assume that there is a partial order compatible with this graph, i.e., an edge between two states exist if one is an immediate successor of the other. Then any degree one vertex is either a local maximum or a local minimum. Since this graph has three degree 1 vertices and in a finite lattice there is only one local maximum and only one local minimum, this particular graph can not be a lattice. □

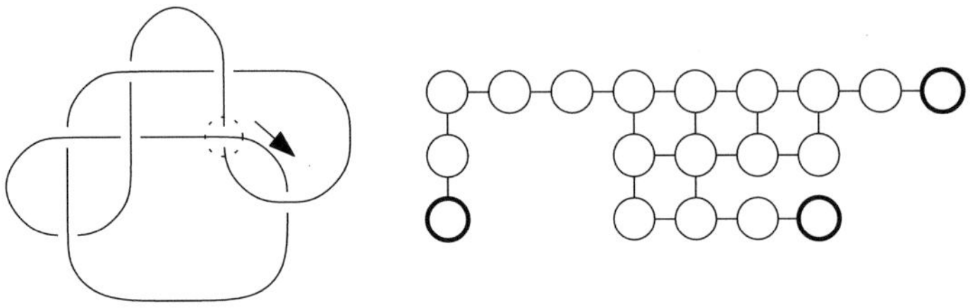

FIGURE 3. The knot 7_6, the chosen root, and its space of states.

3. Proof of Theorem 1.1

In this section we will assume that K is a prime knot and D is a reduced alternating projection for K. Choose a root vertex v_0 in $G(D)$. We will provide an algorithm to go from one rooted spanning tree T_1 to another rooted spanning tree T_2 through a sequence of edge exchanges. We will label vertices v of $G(D)$ with distinct integers.

An initial segment $IS(v,T)$ of a rooted spanning tree T is the sequence of vertices on the unique rooted path from the root to v in T. For $v \neq v_0$, let $\phi(v,T)$ denote the vertex that points to v through an edge not in T. Let $Bel(v,T)$ be a small neighborhood of the set of vertices below v in T, i.e., those that can be reached from v via directed paths in T, the edges between them (not necessarily in T) and the elementary regions surrounded by those edges. Let $Bel_1(w,T)$ be the connected component of $Bel(w,T)$ containing the successor of w in T with the smaller label. When the tree T is obvious from the context, we will suppress T from these notations. The rooted meet of two rooted trees T_1 and T_2 is the connected component of the root in $T_1 \cap T_2$ and will be denoted by $T_1 \cap_r T_2$.

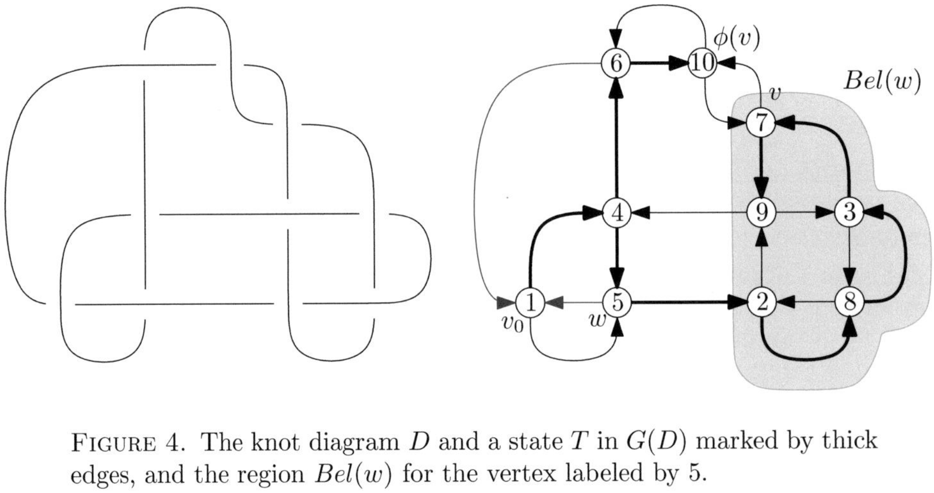

FIGURE 4. The knot diagram D and a state T in $G(D)$ marked by thick edges, and the region $Bel(w)$ for the vertex labeled by 5.

In order to prove Theorem 1.1, we will show that for any given two states T_1 and T_2, we can persistently enlarge $T_1 \cap_r T_2$.

Lemma 3.1. *Given a state T and a vertex w other than the root v_0, if $\phi(w,T)$ is not in $Bel(w,T)$, then there is a sequence of edge exchanges that converts the incoming edge for w into a terminal edge, removing edges only below w.*

Lemma 3.2. *Under the conditions of Lemma 3.1, there is a vertex $w' \in Bel(w,T)$ with $\phi(w') \notin Bel(w,T)$.*

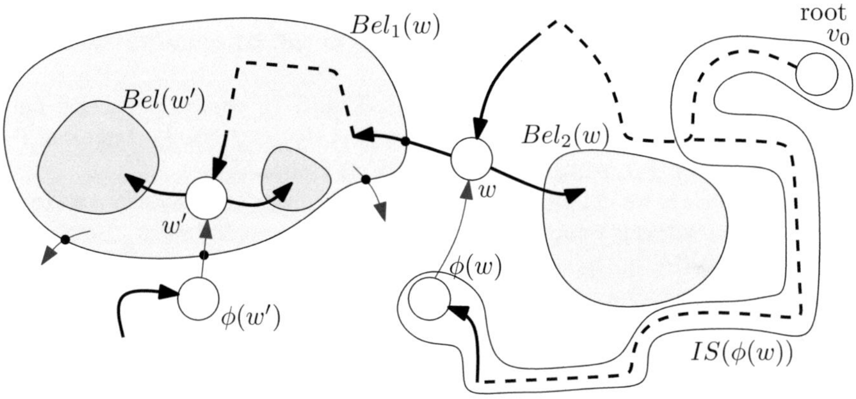

FIGURE 5. Thick edges belong to the spanning tree T.

Proof of Lemma 3.2. Assume $Bel(w)$ is nonempty. Given w, consider $D \cap \partial Bel_1(w)$. Since K is prime and D is reduced, $\partial Bel_1(w)$ is not a separating circle, hence there are at least 4 intersections. Since the orientations of adjacent regions alternate, we get at least two edges entering into $Bel_1(w)$.

Pick a vertex w' among all terminal vertices of edges in G entering into $Bel_1(w)$ not originating from w. This choice implies that $\phi(w') \notin Bel_1(w)$. Furthermore the union $IS(w) \cup \{w \to \phi(w)\} \cup IS(\phi(w))$ contains an unoriented circuit of edges and vertices that separate $Bel_1(w)$ from $Bel_2(w)$ (see Figure 5), hence going from w' to a vertex in $Bel_2(w)$ would take at least two edge exchanges. We conclude that $\phi(w') \notin Bel_2(w)$ as well. $\square$

Proof of Lemma 3.1. If $Bel(w)$ is empty, then w is already a terminal vertex. Otherwise, we will use induction on the depth d of the tree $Bel(w)$.

For $d = 1$, $Bel(w)$ could contain up to two vertices. If there is only one vertex, it is a terminal vertex and an edge exchange empties $Bel(w)$. If there are two vertices, $Bel(w)$ has two components, which as in the proof of Lemma 3.2, are not adjacent to one another.

Lemma 3.2 tells that an edge exchange at either vertex decreases the size of $Bel(w)$, and we are led to the case of one vertex.

Assume the hypothesis is true for all trees $Bel(w)$ of depth d and less. If $Bel(w)$ has depth $d+1$, use Lemma 3.2 to find a w' with the property $\phi(w') \notin Bel(w)$, in particular $\phi(w') \notin Bel(w')$. Therefore by the induction hypothesis w' becomes a terminal vertex after a finite sequence of edge exchanges only removing edges in $Bel(w')$. Then performing an edge exchange at w' decreases the size of $Bel(w)$. Hence repeating this process w becomes a terminal vertex while only edges below w being removed throughout the process. □

Proof of Theorem 1.1. Given two distinct rooted spanning trees T_1 and T_2, pick a vertex w adjacent to $T_1 \cap_r T_2$ along an edge in T_2. Note that $w \neq v_0$ since $v_0 \in T_1 \cap_r T_2$ and once w is a terminal edge, after an edge exchange w will be connected to v_0 along the same rooted Hamiltonian path as in T_2.

Let v_1, v_2 be the vertices that lead to w in T_1 and T_2 respectively. By definition, $v_2 = \phi(w, T_1)$ and $v_2 \in T_1 \cap_r T_2$. Hence $IS(\phi(w,T_1),T_1) \subset T_1 \cap_r T_2$, but $w \notin T_1 \cap_r T_2$, hence $w \notin IS(\phi(w,T_1),T_1)$, which means $\phi(w,T_1) \notin Bel(w,T_1)$.

Applying Lemma 3.1, we get a sequence of edge exchanges that ends in a state where w is a terminal vertex without removing any edges from the rooted meet. Now perform an edge exchange at w, this enlarges the rooted meet. Since the rooted meet only enlarges during this process, in finitely many repetitions of this process we reach T_2. □

4. An application to $(2, 2n+1)$ torus knots

In this section we will provide a different proof of the following result originally proved by Ozsváth and Szabó:

Theorem 4.1. *The $(2, 2n+1)$ torus knots are characterized among alternating knots by the Alexander polynomial.*

Proof. Let D be a reduced alternating projection for a knot K with Alexander polynomial $\Delta_K(t) = 1 + (-t) + (-t)^2 + ... + (-t)^{2n}$. Since all coefficients of powers of $-t$ are $+1$, each state has a different weight and $\Delta_K(t)$ is not a product of two alternating knot polynomials (c.f. [OS05, Prop. 4.1]), hence K is prime.

Let T_0 be the state with the least t power. Since K is prime and the fact that each edge exchange changes the power of $-t$ by ± 1, using Theorem 1.1 we get a linear ordering on the $2n+1$ states starting at T_0, reaching each next state by exchanging an edge of weight $+1$ with an edge of weight $-t$.

According to Proposition 2.5 and due to this linear order, T_0 and the top state T_{2n} have only one terminal edge each, hence they have no branching, whereas intermediate states have 2 terminal edges.

Since Proposition 2.4 tells that each edge v (except the two that point to v_0) can be extended to a state having v as a terminal edge, and since we can reach that state from T_0 by positive edge exchanges, we see that all edges in T_0 have weight $+1$. We conclude that T_0 has $2n$ edges since each edge of weight $+1$ is used only once in an edge exchange and no new edges emerge with weight $+1$ as we go from T_0 to T_{2n}.

Edge orientations and weights do not depend on the choice of the root vertex, hence, after moving the root from v_0 to v_1, we still get a space of states with the same properties, in particular, there will be a new state T_0' containing a linear directed chain of $2n$ vertices starting at v_1, ending at v_0. Hence we get a cycle of length $2n+1$ of edges of weight $+1$. Similarly, all remaining edges have weight $-t$, form a cycle and are used in T_{2n}, except the one pointing at the root.

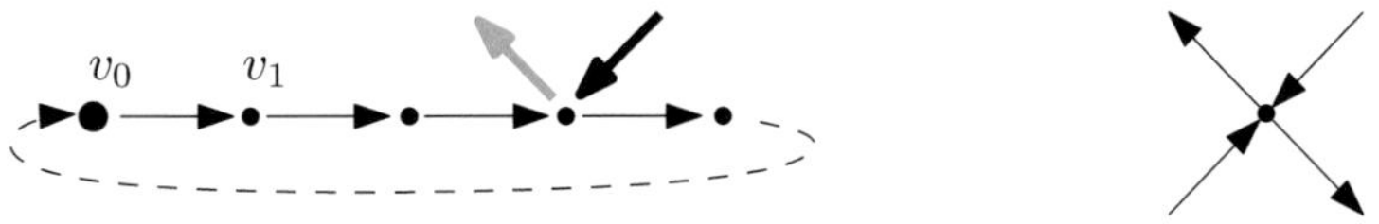

FIGURE 6. Thick edges have weight $-t$. On the right, orientations at a typical node.

This information tells us that if there is an incoming edge of weight $-t$ at a vertex v, the next edge of weight $-t$ has to be on the same side of the cycle of $+1$ edges due to the cyclic alternating orientation of edges at a vertex. Since these edges with weight $-t$ form a cycle as well, they have to go between consecutive vertices. This gives us the diagram for the $(2, 2n+1)$ torus knot. □

Acknowledgements. I would like to thank Bedia Akyar for the invitation to give a talk at Dokuz Eylül University, during which time period I started exploring the properties of the Crowell state space. Most of this work was done during my time at Ferris State University. I would also like to thank Mahir Bilen Can and Mohan Bhupal for providing inspiring feedback.

References

[Cro59] R. H. Crowell, Genus of alternating link types, *Ann. of Math.* (2) **69** (1959), 258–275.

[GL86] P. M. Gilmer, R. A. Litherland, The duality conjecture in formal knot theory, *Osaka J. Math.* **23** (1986), 229–247.

[OS05] P. Ozsváth and Z. Szabó, On knot Floer homology and lens space surgeries, *Topology* **44** (2005), 1281–1300.

[Kau83] L. Kauffman, *Formal knot theory*, Mathematical Notes, Vol. 30, Princeton University Press, 1983.

MICHIGAN STATE UNIVERSITY
E-mail address: durusoyd@math.msu.edu

19th GÖKOVA GEOMETRY-TOPOLOGY CONFERENCE
May 28 – June 2, 2012

List of Participants

Acu, Bahar	University of Southern California, USA
Akbulut, Selman	Michigan State University, USA
Akduman, Setenay	Dokuz Eylül University, Turkey
Argüz, N. Hülya	University of Hamburg, Germany
Arıkan, Fırat	Max Planck Institute for Mathematics, Germany
Aydın, Emel	Yaşar University, Turkey
Baykur, İnanç	Max Planck Institute for Mathematics, Germany
Beyaz, Ahmet	Middle East Technical University, Turkey
Bayındır, Haldun Özgür	Middle East Technical University, Turkey
Behrens, Stefan	Max Planck Institute for Mathematics, Germany
Bhupal, Mohan	Middle East Technical University, Turkey
Biçer, Gülsüm	Aksaray University, Turkey
Bloom, Jonathan	Massachusetts Institute of Technology, USA
Brugallé, Erwan	Institut Mathematiques de Jussieu, France
Can, Mahir Bilen	Tulane University, USA
Cengiz, Mustafa	Boğaziçi University, Turkey
Coşkunüzer, Barış	Koç University, Turkey
Çelik, Münevver	Middle East Technical University, Turkey
Çelik, Fatih	Koç University, Turkey
Çengel, Adalet	Middle East Technical University, Turkey
Çevik, Gülen	Koç University, Turkey
Çoban, Hatice	Middle East Technical University, Turkey
Dalyan, Elif	Hitit University, Turkey
Demir, A. Sait	Istanbul Technical University, Turkey
Durusoy, Daniel Selahi	Michigan State University, USA
Deniz, Deniz Pınar	Ankara University, Turkey
Eden, Sinan	Instituto Superior Tecnico, Portugal
Ege, Özgür	Ege University, Turkey

Eğilmez, İlknur	Koç University, Turkey
Erdal, Mehmet Akif	Bilkent University, Turkey
Erden, Meltem	Ege University, Turkey
Erdoğan Demir, Sultan	Bilkent University, Turkey
Etgü, Tolga	Koç University, Turkey
Finashin, Sergey	Middle East Technical University, Turkey
Genç, Özhan	Middle East Technical University, Turkey
Güçlükan İlhan, Aslı	Bilkent University, Turkey
Güğümcü, Neslihan	Koç University, Turkey
Güneş, Çisem	Bilkent University, Turkey
Hayano, Kenta	Osaka University, Japan
İşal, Hanife	Middle East Technical University, Turkey
Kalafat, Mustafa	Middle East Technical University, Turkey
Karaca, Bahriye	Yeditepe University, Turkey
Karakurt, Çağrı	University of Texas, USA
Karatuğ, Ozan Bircan	Koç University, Turkey
Kartal, Yusuf Barış	Middle East Technical University, Turkey
Kaya, Elif Tuğçe	Ege University, Turkey
Kirby, Rob	UC Berkeley, USA
Kişisel, Özgür	METU North Cyprus Campus, North Cyprus
Kocaayan, Hicran	Ege University, Turkey
Korkmaz, Mustafa	Middle East Technical University, Turkey
Koçak, Şahin	Anadolu University, Turkey
Kutluhan, Çağatay,	Harvard University, USA
Leung, Conan	Chinese University of Hong Kong, Hong Kong
Lekili, Yankı	University of Cambridge, UK
Lipyanskiy, Maksim	University of Columbia, USA
Manolescu, Ciprian	UC Los Angeles, USA
Medetoğulları, Elif	Atılım University, Turkey
Mehreliyev, Tevekkul	Koç University, Turkey
Mikhalkin, Grigory	Université de Genève, Switzerland
Neşe, Hüden	Boğaziçi University, Turkey
Okutan, Osman Berat	Bilkent University, Turkey
Önder, Turgut	Middle East Technical University, Turkey

Özbağcı, Burak	Koç University, Turkey
Özcan, Abdullah Fatih	İnonü University, Turkey
Özer, Arda Buğra	METU North Cyprus Campus, North Cyprus
Özgüner Akyol, Ayşegül	Bilkent University, Turkey
Öztürk, Ferit	Boğaziçi University, Turkey
Pamuk, Mehmetcik	Middle East Technical University, Turkey
Pamuk, Semra	Middle East Technical University, Turkey
Petrunin, Anton	Pennsylvania State University, USA
Risler, Jean-Jacques	Institut de Mathématiques de Jussieu, France
Sağlam, K. Nur	University of Minnesota, USA
Sakallı, Sümeyra	Koç University, Turkey
Salur, Sema	University of Rochester, USA
Sarıoğlu, Celal Cem	Dokuz Eylül University, Turkey
Seyhun, Merve	Koç University, Turkey
Sivek, Steven	Harvard University, USA
Şavklı, Ece Gülşah	Bülent Ecevit University, Turkey
Şentürk, Berrin	Bilkent University, Turkey
Şimşir, Muazzez	Hitit University, Turkey
Topkara, Mustafa	Mimar Sinan GS University, Turkey
Tunçer, Ata	Koç University, Turkey
Uğurlu, Özlem	Dokuz Eylül University, Turkey
Uğuz, Selman	Harran University, Turkey
Uyanık, Çağlar	University of Illinois of Urbana-Champaign, USA
Ünal, İbrahim	METU North Cyprus Campus, North Cyprus
Varolgüneş, Umut	Massachusetts Institute of Technology, USA
Walter, Benjamin	METU North Cyprus Campus, North Cyprus
Williams, Jonathan	University of Georgia, USA
Yaşar, Fırat	Koç University, Turkey
Yıldız, Eylem Zeliha	Dokuz Eylül University, Turkey
Yiğit, Uğur	Ege University, Turkey
Zabun, R. Arzu	Middle East Technical University, Turkey